THE VOLCK ANTI-UNION ARTWORK

The below early Civil War sketch by Adalbert Johann Volck was done far inside Union territory by a most unlikely Confederate who effectively used pen and ink rather than the sword. For nearly three years while practicing dentistry in Baltimore Volck practiced pen and ink treason against just about anything related to Federal forces opposing the Confederacy. They were so obviously and powerfully anti-Union that Volck was imprisoned by General Benjamin Butler at Fort McHenry early in the war. Volck was a brilliant, well educated German immigrant who fled Bavaria because he had been involved in the 1848 students' rebellion against an absolutist monarchy that ruled in Bavaria and most other European nations. How this European liberal reversed his politics and philosophy to become totally appalled at the thought of ending slavery remains a mystery. His many cartoon caricatures were especially hateful and maliciously directed toward vilification of Abraham Lincoln and his top civil and military staff members.

Volck was one of the Maryland medical professionals who sparked movement of medicines and other critically needed hospital supplies into Virginia in small boats plying the Potomac -- often under the watchful, wary eye of Thomas A. Jones at Popes Creek about 35 river miles south of Washington. The artist himself made a couple of boat trips to Rebel territory up river from the capital, once landing near Leesburg at Ball's Bluff where he met General Thomas J. Jackson leading a group of his staff in prayer. Volck sketched this rare scene and included it in his group of 29 prints that constituted the body of his most memorable work on behalf of the Southern cause.

After the war Volck mellowed and adjusted to its outcome. From then until his death in 1912 he exercised at their fullest his immense talents as painter, etcher and he even excellently executed artwork of low-relief in silver and bronze. Perhaps one of Volck's most notable achievements was his portrait in oil of Robert E. Lee about six months before the general's death.

Adalbert Johann Volck's sketch showing the Potomac River scene repeated countless times during the early months of the Civil War - pro-Confederates on their way to join the Confederate Army in Virginia.

THOMAS A. JONES

CHIEF AGENT OF THE CONFEDERATE SECRET SERVICE IN MARYLAND

by

John M. and Roberta J. Wearmouth

HERITAGE BOOKS
2013

HERITAGE BOOKS
AN IMPRINT OF HERITAGE BOOKS, INC.

Books, CDs, and more—Worldwide

For our listing of thousands of titles see our website
at
www.HeritageBooks.com

Published 2013 by
HERITAGE BOOKS, INC.
Publishing Division
100 Railroad Ave. #104
Westminster, Maryland 21157

Copyright © 1995 John M. and Roberta J. Wearmouth

Other Heritage Books by John M. Wearmouth:

The Cornwell Chronicles: Tales of an American Life on the Erie Canal, Building Chicago, in the Volunteer Civil War Western Army, on the Farm, in a Country Store

Other Heritage Books by Roberta J. Wearmouth:

Abstracts from the Port Tobacco Times and Charles County Advertiser*:*
Volume 1, 1844–1854
Volume 2, 1855–1869
Volume 3, 1870–1875
Volume 4, 1876–1884
Volume 5, 1885–1893
Volume 6, 1894–1898
CD: Port Tobacco Times, *Volume 1–6*

All rights reserved. No part of this book may be reproduced or transmitted in any form or by any means, electronic or mechanical, including photocopying, recording or by any information storage and retrieval system without written permission from the author, except for the inclusion of brief quotations in a review.

International Standard Book Numbers
Paperbound: 978-0-7884-5473-8
Clothbound: 978-0-7884-6901-5

Thomas A. Jones

Chief Agent of the

Confederate Secret Service

in Maryland

by

John M. and Roberta J. Wearmouth

John M. Wearmouth

Roberta J. Wearmouth

Copyright 1995

Stones Throw Publishing
Port Tobacco, Maryland 20677
2000

ACKNOWLEDGMENTS

The authors wish to acknowledge with deepest appreciation the support and encouragement received from the gracious people listed below. This work would not have been so interesting had it not been for their sharing of memorabilia and reminiscences.

Rita Cooksey

Helen Jones Horgan

The late Raymond Anthony Jones

Anna Jones Schellin

Irene Penn Quinn

Carolyn Jones Straeter

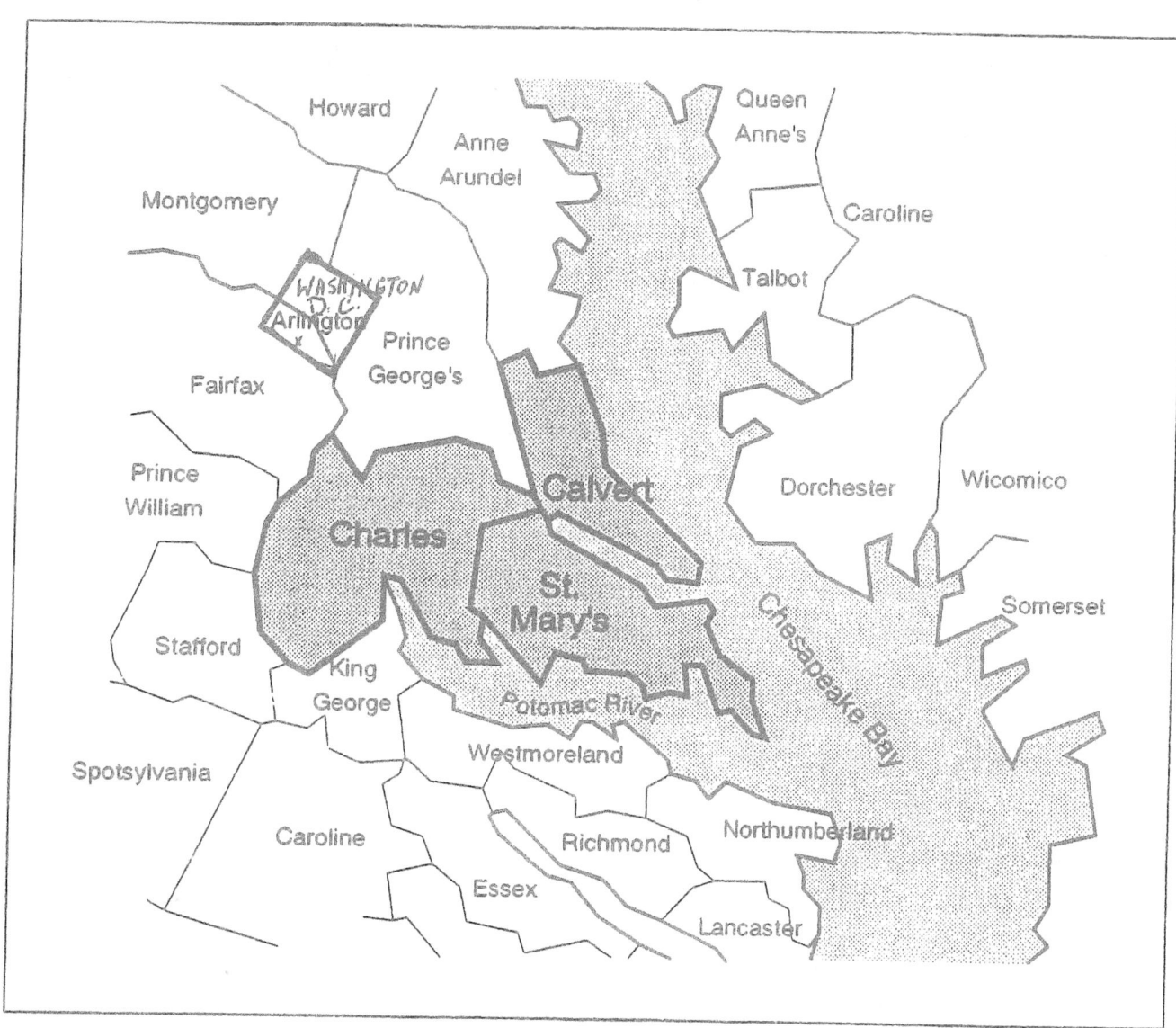

Calvert, Charles and St. Mary's Counties in Southern Maryland

Shaded areas show that part of southern Maryland most devoted to the Confederate cause and most committed to slavery before and during the Civil War.

TABLE OF CONTENTS

	Page
Preface	1
Part 1 Introduction	4
Part 2 Mr. Seward's Heavy Fist	14
Part 3 Under the Federal Presence	26
Part 4 GATH at Center Stage	45
"How Wilkes Booth Crossed the Potomac"	56
By George Alfred Townsend	
Part 5 Introduction to Book	81
"J. Wilkes Booth"	86
By Thomas A. Jones	
Part 6 Return to Charles County	141
Part 7 Smoot's Great Sturdy Boat	149
Part 8 Reflections	154
After Note	160
Addendum A - Genealogical Notes	162
Addendum B - Jones Family News Items	166
Addendum C - Obituary of Thomas A. Jones	168
Addendum D - Appointment of Rev. Lemuel Wilmer to Post Chaplain for the Post of Port Tobacco	170
Addendum E - Oath of Allegiance Signed by Rev. Lemuel Wilmer	171
Bibliography	172

LIST OF ILLUSTRATIONS

Page

Confederates crossing the Potomac	Inside front cover
Maryland map showing areas most devoted to Confederate cause	iii
Jones's home in 1861	3
Photograph of "Cliffton" c.1980	6
Miss Olivia Floyd of "Rose Hill"	9
View of Virginia shore from Jones's porch	12
Adalbert Johann Volck's sketch showing contraband destined for Confederate use	13
Rev. Lemuel Wilmer	15
Thomas A. Jones's Oath of Allegiance	24
Old Capitol Prison in Washington, D. C.	25
Winter scene in Charles County camp of 8th New Jersey Volunteers	27
Navy's first aircraft carrier anchored in mouth of Mattawoman Creek	29
Guard mount in camp of 1st Massachusetts Volunteers camped near Budd's Ferry	31
Thomas A. Jones about 1875	39
Reconstructed 1819 Charles County Courthouse	42
"Prospect Hill," home of George Dent on Pope's Creek Road	43
"Ellenborough," home of Thomas David Stone on Pope's Creek Road	44
Honorable Frederick Stone of Charles County	46
George Alfred Townsend, Samuel L. Clemens, and David Gray	55
"Huckleberry," residence of Thomas A. Jones near Pope's Creek	61
Map showing Booth and Herold escape route	68
Thomas A. Jones about 1890	74
Promotion copy for reprint of *J. Wilkes Booth*	83
Roadside historic event signs for Dent's Meadow and "Rich Hill"	85
Mary Swann Kelly	142
Henry Woodland	148
Barnes-Compton house in Port Tobacco	152
John H. Surratt story	153
William Williams, Federal detective	157
Post Office building at Bel Alton (Cox's Station)	159
"Lincoln 1865"	161
1861 map of the war along the Potomac that year	176-177

THOMAS A. JONES

CHIEF AGENT OF THE
CONFEDERATE SECRET SERVICE IN MARYLAND

PREFACE

During the American Civil War it was often hazardous to be an avowed Confederate sympathizer in any State outside the Confederacy. In southern Maryland, traditional States' Rights and plantation country, which begins immediately south of Washington, D.C., a Confederate's life was always fraught with tension and sometimes with terror. Union cavalry usually could be found within an hour's ride from many locations hereabouts. Yet, a stalwart, known southern Maryland Confederate sympathizer in Charles County (thirty miles from the Federal District), during a few unforgettable days in April 1865 was the king-pin in one of the most intensive, desperate man-hunts ever known in North America. For almost a week the times, places and persons at center-stage of the Lincoln assassination act fed the flames and passions of a frantic southern Maryland search for the murderers of Abraham Lincoln. During all this, Thomas Austin Jones of Pope's Creek at the Potomac River in Charles County quietly and effectively thwarted every effort of the Federal establishment to locate fugitives Booth and Herold. They remained hidden restlessly for five days in a dense growth of scrub pine less than half a mile east of today's busy U. S. Route 301, within the limits of the 20th century village of Bel Alton, about four miles south of the present county seat of La Plata.

Americans have done a respectable job through the years since 1865 of recognizing and recording the devotion and courage of both Confederate and Union heroes. A few of them, however, to this day have not received the acknowledgment and respect they earned during the war. Considering now what was known about Jones's contribution to the Southern cause it may not be too wild a statement to say that what he accomplished during his five years of steady, risky pro-Confederacy secret service would have equaled in value the military contributions of some of Lee's most capable field commanders. Jones's services from his Pope's Creek headquarters were given almost without interruption or respite during the entire course of the conflict. His work was stopped for a short period while he was incarcerated "on suspicion" in Federal prison in Washington, D. C. from October 1861 to March 1862. Miraculously, he was released without conviction and sentencing.

In stage setting for what follows a definition of "Pope's Creek" is in order since most of the Civil War Confederate trans-Potomac River traffic centered on a Maryland neighborhood described somewhat loosely as "Pope's Creek." The very short stream that gives its name to the neighborhood is little more than half a mile long. Over a nearly 400-year period its meandering mouth, subject to constant tidal influences, likely has shifted as much as several hundred feet from today's location. The stream flows through a shallow, narrow valley that is half-saucer shaped and formed by a deep landward indentation of bluffs that bend back from the river on both sides of the creek. The low-land area is primarily marsh that protects and nourishes many varieties of fish and birds. For centuries it has been a haven and rich feeding trough for man, fowl and marine life. Until very recently the Potomac offshore here has been a sea food locker for great annual harvesting of oysters and the blue crab. Environmental changes have altered this scene tragically in late 20th Century.

The name Pope's Creek doubtless derives from an early planter named Francis Pope who purchased a large holding just south of the creek mouth in 1665. This parcel of 350 acres seems to have been the very same bit of real estate purchased by Thomas A. Jones and a few friends in the early 1850's. At this time the land was called "Pope's Creek Farm." For about three centuries Pope's Creek was prominent in agriculture, ferry services, steamboat landings and railroad traffic. Certainly none of these equaled the Civil War importance of the neighborhood as the spout of a funnel which helped the Confederacy keep in touch with the rest of the world.

With the 20th century increase in importance of highway transportation and the construction of the Harry W. Nice Memorial Bridge and subsequent completion of north-south U. S. Route 301 both steamboat and rail traffic at Pope's Creek evaporated. But pumpkins have turned into gustatorial emporiums of seafood delights that have since World War II brought a renewed profile to old Pope's Creek. On most summer weekend evenings as many as a thousand people may be seated in three restaurants constructed on the old beaches and out over the Potomac.

Pope's Creek's commercial core lies about 45 miles south of Washington--nearly 55 by water. It is close to 65 miles south of Baltimore center and nearly 75 miles north of downtown Richmond. Geographically this neighborhood was ideally situated for channeling wartime North-South communications. The creek mouth is very nearly halfway between the Capital and the Chesapeake Bay. Perhaps as important as anything the Potomac was only two miles wide here.

Research about the life of Thomas A. Jones continued for nearly a decade. Most valuable was a mid-1994 meeting with a great granddaughter whose family papers contributed rather substantially to the information in the following work, especially regarding the role of George A. Town send (GATH). During the past seven or eight years, the authors have conducted four taped and three untaped interviews with descendants of Thomas A. Jones. These family members knew surprisingly little about the work of Jones during the war. Probably this resulted from the desire of the immediate family to refrain from calling attention in the 19th century to 20th century family support for the Confederate cause. So, much family information known to the many children of Jones simply did not get handed down to succeeding generations. The past in quiet, tidewater southern Maryland often was thought best comfortably left where and as it had been--shaped by times, causes and bitter losses few can understand today.

Much thought was given to titling this work. The name Thomas A. Jones today means very little even in Charles County, Maryland. Jones's 1890 application for membership in the Society of the Army and Navy of the Confederate States explicitly gave his CSA service rank as *Chief Agent of the Secret Service in Maryland*. And his old commander, Colonel William Norris, Chief of the CSA Signal Bureau, confirmed this title and rank in a hand-written statement on the

Jones' membership application form. This became the basis for the title of this work, which admittedly is primarily about the life of Thomas A. Jones of Charles County, Maryland, 1820-95.

Jones's home in 1861 on bluff overlooking Potomac and facing the Virginia shore. Section on left is the original c.1830 structure in which the Jones family lived from 1858-1863.

PART 1

INTRODUCTION

Of all the names that now may come to mind in connection with Confederacy Civil War accounts, the name Thomas Austin Jones certainly would be among the least known. And yet the Confederate cause had no more loyal, avid and effective single advocate than this quiet southern Maryland farmer whose face, home and heart looked southward toward the Confederacy throughout the war. [1]

Partly because of his home's location, high on a bluff above the Potomac River near the mouth of Pope's Creek in Charles County, Maryland, and in part because of his known, acute feeling of loyalty toward the Southern cause, Jones was recruited very early in the war to work within the Confederate States' secret service.

Little is known about the roots of this Jones family in this part of Maryland. A wealthy, plantation owning Jones family had been prominent in Charles County during the 18th century, but this Thomas A. was not of that privileged family. A family genealogist of early 20th century traced the parentage of Thomas A. Jones to early 19th century newcomers from Wales. Thomas probably had four sisters. The youngest was Ann Mary who was born in 1838 near Port Tobacco and worked there many years as a seamstress. On January 14, 1907 she entered the Home of The Little Sisters of the Poor in Baltimore, Maryland as a resident (not as a member of the Community). She died there on April 8, 1907. [2]

Reasonably sound documentation about the parentage of Thomas A. is in the 1830 census for Maryland. A family of Elisha Jones is recorded as living in the Allen's Fresh District of Charles County. The family included one male under 5 years, one male 5-10 years, (doubtless Thomas A.) and Elisha between 40-50 years, 2 females under 5 years, one female between 30-40 years (Mary Stuart). There were two male slaves; a child and an adult. Thomas A. Jones's obituary of March 1895 mentions a married sister, Mrs. Mary Hindle, who probably was the Mary Ann Jones who died in 1907 while with the Little Sisters of the Poor. Thomas A. Jones was born in Charles County on October 2, 1820. His wife Jane Harbin was born near Bryantown in the same county July 2, 1819. They were married at the Mattawoman Parish Catholic Church (now

---

[1] Jones seems to have been engaged in more than farming before the war. The U. S. Census of 1850, Allen's Fresh District, Charles County, Maryland, lists Jones as a Collector of Taxes, with three children and wife in same household.

[2] Letter from Provincial Residence, Little Sisters of the Poor, Baltimore, Maryland to author, September 2, 1995. This letter stated that Ann Jones was the daughter of Elia (Elisha) and Mary Stuart Jones of Charles County, Maryland, and had been married.

called St. Mary's) at Bryantown January 8, 1845. [3]

The Jones' farm was bounded on the west by the Potomac and on the north by Pope's Creek. Standing in his yard Jones could see as far as seven or eight miles westward up-river toward the mouth of Nanjemoy Creek and down-river to the Virginia terminus of Laidler's ferry crossing, which was anchored at Morgantown near Lower Cedar Point. Here the river was relatively narrow, only about two miles across, and visual communication shore-to-shore was easy night or day under clear weather conditions. Also, the riverside here offered excellent opportunities for landing and hiding boats and good places to land or launch. Added to these advantages was the fact that most of Jones's neighbors were in hearty sympathy with the South, and in fact more than a few of them eluded Union military and naval units and escaped across the river to become active combatants in CSA military forces. On the Maryland side, Jones was ably supported initially be Major Roderick G. Watson whose home "Clifton" lay about 300 yards south of the Jones farm. [4]

Jones himself regularly crossed the river during the war years and was never apprehended while en route. In fact, the "Jones Ferry" carried hundreds of passengers back and forth during the war years, nearly all having some business with the Confederacy. Directly across the river was the farm of Benjamin Grimes (sometimes Grymes) in King George County. He anchored the Confederate Potomac communication link on the Virginia shore. The Federal government was fully aware of its near helplessness when it came to stopping or even seriously hindering this trans-Potomac River ferry operation. Throughout the war the Federal navy patrolled the river assiduously and Union Cavalry patrols guarded the Maryland shore from Fort Washington to Point Lookout. Regardless, the trans-Potomac ferry-boat linkage continued to function very effectively. The great Union Navy gill net mesh was too coarse to trap the smaller ferry fish like Jones.

Signaling from the Maryland side was done from a second floor bedroom window of Watson's home. This house was destroyed about 1980 by vandals. Boats used for the Jones-Grimes mail service were fast and sturdy, and were of course kept on the more secure Virginia side. Then, as now, just before sunset on any clear day the reflections of the high bluffs near Pope's Creek fall out into the Potomac until they nearly meet the shadow cast by woods on the Virginia side. At that time it's very difficult to locate small craft out on the river. So boats from the Virginia side usually crossed just before sunset, leaving their packages and letters from Richmond in the fork of a dead tree on Jones's property. Often mail would be there waiting to be taken south if Jones himself was not there to meet the Virginia mail. When Marylanders felt it

---

[3] Hall of Records, State of Maryland, Annapolis, Maryland. Charles County, Maryland, Bryantown/Mattawoman Church records, film #0013759,

[4] Roberta J. Wearmouth, *Abstracts from the Port Tobacco Times and the Charles County Advertiser, Volume Two: 1855-1868*, Heritage Books, Bowie, Md.,1991. (Hereinafter *Abstracts*). The January 2, 1862 issue announced the death of Watson at his home on November 8, 1861. He was 56 years old, ". . . a man of most kindly, generous, affable disposition . . . kind husband, father and master."

wasn't safe for the Virginia boat to set out, a black dress or shawl was hung in a dormer window of the Watson house. At night light signals were used. A very sophisticated light-reflector arrangement was uncovered at the Grimes house by a Yankee patrol in 1861. Usually the signaling device was handled by Roderick Watson's daughter, Mary. After Mr. Watson died Mary carried out the signaling operation herself until the war ended. Jones once wrote about Mary:

> *"Miss Watson was a remarkably pretty young lady, about 24 years of age. She would have made almost any sacrifice for the Confederacy, and I know that I owe in great measure the success which attended the management of the Confederate mail to her ceaseless vigilance and skill."*

After the war Miss Watson married a Confederate blockade runner and moved to California.

Overlooking the Potomac River "Cliffton" was the home of Roderick G. Watson who operated the signal station until his death about the year after the war began. After that, his daughter, Mary, continued the signal service. The house was about 1/4 mile south of the Jones residence.

So, Jones continued his extremely valuable service on behalf of the Confederacy without detection throughout the war. With notable help from a few other Charles County personages including Dr. Stoughton W. Dent of "Dent's Palace" at Centreville (now Dentsville); George Dent, Sr. who lived at "Prospect Hill" (now called "Keechland") at Pope's Creek; Thomas H. Harbin, who probably lived near Bryantown, about twenty miles east of the river.

Another, lesser known figure, was James Alexander Brawner, owner and proprietor of the Brawner House in Port Tobacco, which on any given evening during the war probably sheltered more Confederate sympathizers than any hotel of like size in Virginia. The June 18, 1886 edition of the *Port Tobacco Times and Charles County Advertiser* announced the death of J. Alexander Brawner and recalled that he had been

> "*...a strong sympathizer of the Southern cause and his hotel was headquarters for leading spirits* [of all types] *among those sympathizers in our county. By his effort and energy many of our citizens who entered the Confederate army were enabled to cross the river and reach Southern lines, and more than once he was suspected of this by the Union authorities here and his person and hotel made the subject of especial espionage in consequence but they were never able to attach the blame to him and he was never molested.*"

Perhaps the most important of the mail carrier assistants was Dr. Dent, who simply because he was a doctor was able to carry unnoticed in his voluminous duster cover-all many letters and even small quantities of medicines desperately needed across the Potomac. Chloroform, for example, was delivered in Virginia in jugs labeled "Neats-foot oil." [5]

Miss Olivia

One southern Maryland Civil War period figure may have descended wraith-like through the years to become better known in some ways than most of the principal role players in the Booth-Jones affair. Contemporary official Government documents say nothing about Miss Olivia Floyd of "Rose Hill" at Port Tobacco, alleged to have done espionage work for the South. Local documents and publications and writing of local Confederate sympathizer-contemporaries tell us nothing substantial about Miss Floyd. Chivalry still existed in isolated southern Maryland pockets during the war and it may be that at least the knowledgeable gentlemen chose not to involve one of their very own ladies in the sometimes messy tangle of spy work.

---

[5] *The War of the Rebellion: Official Records, Series II - Volume II*, p. 866, Washington: GPO, 1897. Hereinafter cited *W.R.O.R.* Note: The official government documentation for the *W.R.O.R., Series II - Volume II*, "Suspected and Disloyal Persons," is based almost entirely on U.S. National Archives Records in Records Group 59 (RG 59), *General Records of the Department of State, Civil War Papers 1861-65*, "Records of Arrest for Disloyalty - 1861 and 1862." Hereinafter cited *RG 59*.

In 1973 the author taped an interview with one of the then rather ancient Spalding girls from "Pleasant Hill" at Pomfret, who described visiting Miss Floyd at Rose Hill as a very young girl, accompanying her aged aunts from "Green's Inheritance," next door to the venerable St. Joseph's Catholic Church at Pomfret. Her description of Olivia Floyd squares nicely with the circa 1890 photo reprinted here (taken with a very early "Kodack"). [6]

The earliest known detailed information about the nature of Miss Olivia appeared in an article written by Mr. William E. Shields for the *Port Tobacco Times* of May 7, 1897. In most provocative words, he closed in saying Miss Floyd at Mr. Shield's next, and promised visit, would give the writer a full accounting of the local "underground" communication system during the time she was part of the chain...these words were written probably within hours after the Shields-Floyd interview in the old parlor, immediately below the bedroom in which Washington had once slept as a guest of Dr. Gustavus Richard Brown, a man who years later tried to comfort the Great Man as he lay dying at Mount Vernon.

Shields, who wrote as "Judge Leisure," was a Yankee type who soon after the interview went back home to Ohio to edit a newspaper that favored the Democrats. So, he was unencumbered by any family or sentimental ties to the late Confederate cause. But he was obviously overcome by the wit and charm of Miss Floyd.

The "Judge" wrote that...

> *"If the rest of the Confederacy's women were as brave, as bright, as loyal and true to the cause she believed just it was no wonder it required four years for the Northern arms to achieve victory."*

So, either up front, or quietly behind covert scenes, men like Thomas A. Jones knew they could count on such strengths to support them during all the war years. [7]

Perhaps during the unrealized next visit to "Rose Hill" Miss Floyd would have cleared up her alleged involvement in the matter of the twenty Confederate raiders who moved from Canada into Vermont in October 1864 and held up the town of St. Albans...at least all the banks. The raiders were driven out of town and some of them were placed on trial in Canada as part of extradition proceedings that resulted from the United States trying to get at least some of the raiders back so they could be tried as spies. The raid was conducted by escaped Confederate prisoners of war wearing civilian attire and bore no resemblance to a military action. No matter, Miss Floyd played a key role in relaying documents from Richmond to Canada signed by Jefferson Davis that made it incumbent upon the Canadians to consider the raiders to be Confederate

---

[6] John M. Wearmouth taped interview with Mrs. Stephen Latchford in 1971 at her home in Mount Rainier, Maryland.

[7] *Port Tobacco Times and Charles County Advertiser*, Vol. LIII, Number 50 of May 7, 1897.

Miss Olivia Floyd (c.1835-1908) of "Rose Hill" near Port Tobacco. Long thought to be one of southern Maryland's liveliest and most courageous spies during the Civil War. She was never officially accused and lived the remainder of her life in her magnificent colonial home. Photo c.1890 taken with Eastman Kodak round photo film camera. Courtesy National Archives.

Historic American Building Survey Photo by T.T. Waterman, 1937, showing newly renovated "Rose Hill," Port Tobacco. This house is located on Blue Dog Hill. Built by George Washington friend Dr. Gustavus Richard Brown in 1774-5.

soldiers rather than civilian robbers. They were saved by the bell . . . the end of the war. It seems that lawful belligerents in Canada were not subject to extradition law, and could not be tried as criminals there. [8]

Not all ferry services involved materials and matters of mercy and aid to the innocents. A statement in Official Records section about "Prisoners of War" etc., by one George W. Smith of Charles County details the nature of items sent to Pope's Creek for delivery to the Confederacy:

> *"The secession feeling commenced about April last, 1861. The principal leaders in the secession party and those who have aided against the Government are, first, James A. Mudd; lives about one mile from Bryantown; has been conveying men and boxes supposed to contain munitions of war from Baltimore and different counties in the State to Pope's Creek on the Potomac. The men were strangers from Baltimore and other places. Mudd paid the expenses. This has been carried on since April last [1861]. Hillary Burch, at Benedict, has been running a regular line of passengers to Pope's Creek in a wagon, carrying arms, etc. daily to Pope's Creek. Robert L. Burch, of Bryantown, has been carrying arms and ammunition to Pope's Creek in his wagon, driven by Dennis Burch. Y. Posey of Port Tobacco, has been running a regular line from [there] taking passengers, arms and ammunition from Benedict to Pope's Creek. Thomas A. Jones, of Pope's Creek, is the man who receives the men, arms and ammunition at that place and conveys them over to Virginia in his own boat and with his own Negroes. Thomas Stone, at Pope's Creek, is also engaged at the same business. Stone has also been collecting men to carry over to Virginia. Luther Martin, of Allen's Fresh, Charles County, is now daily running men and ammunition from Benedict to Pope's Creek.*
>
> *The boxes containing the guns about a week ago came from Anne Arundel County, Md. Supposes at least 200 have left the neighborhood of Bryantown and been carried to Virginia to the secession army. The wagon that carried the passengers, arms and ammunition from Bryantown to Pope's Creek belongs to the cavalry company at Bryantown and is in possession of James A. Mudd..."*

---

[8] Oscar A. Kinchen, *Women Who Spied For the Blue and Gray*, (Philadelphia: Dorrance & Co., 1972).

[Above testimony recorded by Mr. T.B.Robey] [9]

Thomas Austin Jones sacrificed as much for the Confederate cause as did many high ranking political and military leaders from Confederate states. He lived under constant threat of recognition and apprehension by nearby Federal authorities and had the nature of his operations been completely understood he would probably have been treated as a spy. Dr. Charles E. Taylor stated in his 1903 booklet that "perhaps the most useful of all the men connected with the C. S. Secret Service was Mr. Thomas A. Jones of Maryland." [10]

Jones was not a young man when the war started. He was over 40 years old and the father of eight living children, the oldest being only 15 years of age. His wife, Jane, died January 29, 1863, leaving him with many children. During the war he lost possession of his splendid, highly productive 300-plus acre farm on the level high lands overlooking the Potomac. In 1863 he moved to a smaller farm and dwelling called "Huckleberry," which was located nearly a mile inland from the river but in the same vicinity and east of the adjacent waterfront property of George Dent, Sr. (This house and surrounding property now is part of the Loyola Retreat.)

But Jones's greatest time of trial and testing came after the surrender at Appomattox on April 9, 1865 when he had absolutely nothing to gain and much to lose for any actions taken to support an already dead Confederate cause. An extremely complex combination of circumstances entangled Jones deeply in the overall Abraham Lincoln assassination tragedy. Early Easter Sunday in April 1865 Jones found himself with primary responsibility for the overall safety, welfare and eventual escape of the man who had jumped from the presidential box to the stage of Ford's Theater late in the evening of the 14th.

The assassination of Abraham Lincoln dragged a sleepy, tidewater community in southern Maryland onto the stage of national notoriety. The escape route of John Wilkes Booth from the alley behind Ford's Theater led through two counties of southern Maryland -- Prince George's and Charles. Both were long known to be strongholds of Confederate sympathy. Booth and his companion, David Herold, already had surveyed several neighborhoods between the Anacostia River and points along the Potomac shore in Charles County. They felt sure of finding friends and general support along the roads leading to a point near the mouth of the Port Tobacco River, which seems to have been the general locale from which Booth would begin a Potomac River crossing.

Booth and Herold escape planning turned sour at the outset when the actor apparently injured himself leaping to the stage of Ford's Theater after the fatal Derringer round exploded into the President's head from the rear. Booth's painful injury forced him to seek the earliest possible medical attention, and as all students of Lincoln assassination history and lore know this led

---

[9] *RG 59.*

[10] Dr. Charles E. Taylor, *The Signal and Secret Service of the Confederate States*, 1903.

Booth and Herold toward the home of a Charles County physician whom both of them knew through earlier fleeting associations. Booth and Herold told Dr. Mudd that the injury resulted from Booth's horse having fallen en route from Washington. After Booth's injury was treated by Dr. Samuel A. Mudd at his home and after a night's rest at this place the two fugitives set out on horseback late in the afternoon of April 15th and headed painfully and uncertainly toward a pre-determined point on the Potomac.

During the time following the Booth and Herold departure from the Mudd home to the discovery of the two men by Union troops at the Garrett farm near Port Royal, Virginia April 26th, nothing was known publicly of their whereabouts. And, for about two decades this mystery unsettled many people, but most of all it plagued young journalist George Alfred Townsend. His fascination with the Lincoln assassination story began as Townsend gazed upon the president's remains at rest in the East Room of the White House during the funeral ceremony.

In 1865 Townsend published a booklet describing assassination aftermath events in Washington through the conspirators' executions at Fort McNair. The events were described for the *World* newspaper and printed in *The Life, Crime and Capture of John Wilkes Booth and the Pursuit, Trial and Execution of His Accomplices*. At this time nobody knew how Booth had escaped across the Potomac and precisely when. Only Townsend's discovery of the performance of Jones in this story led to complete revelation about the days in a pine thicket ending in a river jaunt in a plain wood skiff furnished by Jones.

A Jones-eye view of the Virginia shore from the north end of his home 1861-63 - about mid-day in 1995.

Adalbert Johann Volck Civil War sketch showing contraband destined for Confederate use being unloaded on the Virginia shore. Much of this included extremely valuable medical supplies as well as munitions and weapons.

PART 2

MR. SEWARD'S HEAVY FIST

Increasing Washington obsession with Confederate influence in southern Maryland began in late summer of 1861. The July Union rout at Bull Run had unnerved the Capital because General Irvin McDowell's retreat exposed the city's defenses on its most vulnerable approaches - the Virginia roads leading to the Potomac. South side defenses were perhaps weakest. Any threat of a Confederate crossing of the Potomac south of Washington was too terrifying to take seriously. A Rebel army coming through southern Maryland doubtless would enjoy the full support of at least ninety percent of the local people. In fact, many would be expected to join armed Confederate military units as soon as they touched the Maryland shore...from Fort Washington to Point Lookout. This was a distance of about 100 miles and nearly all of it under only the most uncertain observation of a thin, long line of Navy vessels whose primary purpose was to protect communications and help feed reinforcements into the Capital City.

During the month after Bull Run one high Government figure played a key and increasingly important part in assessing the danger of pro-Confederate southern Maryland to the safety of Washington. Since perceived threats from the south side were as yet in late summer basically matters of continuing communication with the Confederate States and the opening and conducting of various avenues of intercourse across the Potomac River. Troop reinforcements, arms, mail, food, medical supplies, and espionage efforts were increasing daily south of Washington. Because there did not appear to be any immediate threat of Rebel military action steps were planned to identify and eliminate what were considered to be treasonous acts and open espionage at many points along the river. In 1861 such matters were held to be the responsibility of the State Department rather than military or naval services.

And top gun at the State Department was William E. Seward, a powerful figure in Mr. Lincoln's new cabinet. Here was a man who smarted with resentment because he felt he had been denied the Presidency. In itself this would doubtless have peeved Seward beyond measure no matter who had been nominated at the "Wigwam" during May 1860 in Chicago. What rankled the Secretary most of all was the fact that a noisy, semi-literate group of western plains bumpkins had totally undermined the once powerful Seward-New York State political machine. The Easterners had gone to the Midwest to participate in the Republican Party convention that picked as party leader one of the most unlikely men ever to run for the Nation's top political office. In fact, Seward felt Lincoln surely was not the man of the moment for what seemed to be a probably brewing onslaught of antagonistic forces that just might be set on track toward a total dissolution of the Union. As many other astute politicians thought during 1860 and 1861 the prairie barrister simply showed very little promise of becoming the type of Presidential timber desperately needed for providing the cement that could stabilize and preserve the government that during the previous "four-score-and-ten" had served the infant republic marvelously well.

So now, as one of the most powerful cabinet members, Seward, lacking confidence in Lincoln's administrative and command talents, took a firm hold of matters directly related to national security. And he did so legitimately because the State Department had power and responsibility for all matters related to treason, espionage, immigration, and Federal pardons. And Mr. Seward was apparently fully aware of how his authority clothed him with the responsibility for examining closely any treasonous activities along the Potomac shores. And operations to fortify the Capital City were far from complete. Events and conditions on landward routes into Washington were a matter of most urgent consequence. Seward's State Department jumped at the opportunity to make its Federal authority felt and feared all along the Potomac on both sides. As a sidelight, anything Seward might do to highlight any real or imagined Lincoln ineptitude would be a plus feature of his Department's forays into the southern Maryland counties.

Long-time Port Tobacco Parish rector, Lemuel Wilmer. Served as official U.S. Army chaplain for Port Tobacco Post. He regularly reported to the U.S. War Department on activities of pro-Confederates in his parish. Served as rector from 1822 until his death in 1869.

Late in the 1861 summer Federal authorities were receiving reports of pro-Confederacy activity in Maryland counties south of Washington. On August 26 a man in Philadelphia informed Postmaster General Montgomery Blair that his wife had been asked to forward letters going south of the Potomac to Dr. and Mrs. Stoughton Dent in care of the Port Tobacco, Maryland post office. At this same time John Atwell of Washington visited his wife's family at Pope's Creek and was briefed by them about the Pope's Creek to Persimmon Point, Virginia ferry service that carried anyone across the river, night or day, no matter what their business. Doubtless unbeknownst to many Charles Countians at this time (1861) the rector of Christ Church, Port Tobacco Parish, in the village of Port Tobacco, hot-bed of Confederate sympathizers, was conducting a bit of pro-Union counter-espionage. The Washington Army provost marshal received from local Reverend J. H. Ryland some information sent from the Reverend Mr. Lemuel Wilmer at Port Tobacco. This good Episcopalian man of the cloth informed his Washington church friend about the Confederate cause leaders at Port Tobacco . . . Mitchells, Stones, Brent, Posey and Middleton and the chief agents at Pope's Creek being George Dent and Tom Jones [Thomas A. Jones]. The Reverend Ryland promptly passed the Wilmer intelligence along to the provost marshal in the city. [11] (See pages 171 and 172 about the Wilmer appointment as Union Port Tobacco Post Chaplain.)

On October 26, E. J. Allen, working for Brigadier General Andrew Porter, drafted a long report of his trip through southern Maryland to study CSA support operations. Following orders by the Washington City Provost Marshal, Allen sent his secret service operatives southward toward Pope's Creek with the express purpose of arresting Jones, George Dent, Sr., and James Grimes in Virginia. On the way Allen's detectives picked up a mounted company of Union troopers commanded by Lieutenant W. M. Wilson of the 4th (Regular) Cavalry, to provide the expedition's muscle. Both Jones and Dent heard about the arrest plan and quickly crossed the river, hiding in short-lived safety on the Virginia shore while their families were left to deal with an unwelcome visit of Yankee soldiers. Lieutenant Wilson thoroughly searched the Jones and Dent houses (about half a mile apart) and confiscated all correspondence related to the trans-Potomac mail service. Odds and ends of personal arms, ammunition and para-military accouterments also fell prey to Northern searchers.

About the 18th of the month one of Allen's men went through Port Tobacco just about the time that a troop of Wilson's cavalry left the village after a brief rest stop. He wrote in his report:

> *"At Port Tobacco, I found a troop of U. S. Cavalry and the people all in a state of the most intense excitement. As soon as the cavalry left, the inhabitants held meetings at the taverns and stores, and throwing off all restraint, talked over their opinions so freely that I could easily see that Union men were almost unknown in that locality."*

---

[11] *W.R.O.R. II-II*, p 862.

The Port Tobacco folk did not know that the Union detective and two companions were not quite "secesh" advocates . . . they had passed the word in town that they were simply looking for a boat to take them to Virginia. And eventually they did go under very peculiar circumstances. The Yankee sleuth fell in with Charles County sheriff John Shackleford's effort to move contraband carbines from the village to Chapel Point on the Potomac where the arms, Shackleford, and the Federal Government men were to meet a boat that would take all of them to King George County. Here the Union men were to investigate the complicity of the Grimes family in the trans-Potomac ferry service and its rather simple but effective shore-to-shore signal service. [12]

As might be expected about this time, the name of the notorious (at least in southern Maryland) Lafayette C. Baker works into the story. No more anti-pathetical Yankee ever threw his weight around in this area with greater enthusiasm in all things related to protecting the Union cause. On November 27, L. C. Baker drafted a lengthy report to the Honorable William H. Seward, which was launched with this preamble:

> *"In compliance with orders issued from your Department under-dated November 18th I repaired to the Head Quarters of Brigadier General Hooker at or near Budd's Ferry and was promptly furnished with one hundred men from the Third Indiana Cavalry, under command of Captain Keister. The object of the expedition was to arrest parties suspected of rendering aid to Virginia Rebels, to discover the channels through which contraband and correspondence was being carried on, and if necessary to take into custody any persons found in arms against the United States Government."*

It is not surprising that Baker's first stop heading southward was old Port Tobacco itself. The village already was recognized by Washington as highly vocal and treasonably and effectively active in the enthusiastic total support of the Rebellion. And Baker knew all about this. His memoirs (1866) indicate a complete uncomplimentary disenchantment with everything having to do with Port Tobacco and its courthouse and the adjacent swamps, which were indeed within feet of the west wall of the local palace of justice . . .the same in which in late 1864 Lieutenant Walter Bowie held hostage a troop of Yankee cavalry just long enough to allow his men to hustle aboard Yankee mounts and head northward toward Annapolis, bent on any kind of mischief their mission might warrant. [13]

---

[12] *RG 59*, October 26, 1861. Rpt. of E.J. Allen in the Case of Dent, Jones, the Grimes, the Watkins, Harbin.

[13] L. C. Baker, *History of the U. S. Secret Service*, Philadelphia: 1866.

During this reconnaissance Baker stopped at Allen's Fresh, Newport, Chaptico, and Leonardtown (the last two towns in St. Mary's County). In each place he found a postmaster who although being paid with Federal dollars, violently opposed the Union and openly supported the Southern cause. Baker found things so overtly treasonous in Leonardtown that he advocated in his report that this town be placed under martial law immediately with a U. S. provost marshal appointed to control its people...all traitors and no Union men. He felt the overland traffic from Millstone Landing at the mouth of the Patuxent to the Potomac crossings made it a most critical place for attention. He uncovered in St. Mary's County a group called the Lower Maryland Vigilance Committee. It operated around the most rabid secessionist group in Maryland . . . E. H. Jones, W. M. Able, E. M. Sissell and R. L. Hayden. According to Baker, E. H. Jones alone sent 400 stands of arms to Virginia very early in the conflict. Able, a doctor, who when arrested by Baker, threatened to kill a Yankee for every day he was imprisoned. The Yankee detective enjoyed no success whatsoever in getting Union men of St. Mary's County to assist in uncovering Confederacy supporters. They told him that as soon as the Yankee troopers left local Confederates would burn their buildings, as threatened, and they were terrified about the perceived certainty of being assassinated or hanged by their neighbor-members in the Vigilance Committee.

Baker reported to the Postmaster General on the situations found in the southern Maryland post offices with respect to disloyalty. He was ordered by Montgomery Blair to displace all treasonous postmasters on the next foray south and replace them with trustworthy Unionists where possible. If suitable replacements could not be found the post offices were to be closed. [14]

The neat, comfortable world of Thomas Austin Jones tumbled down about him when he was arrested by a Union Army attachment of General Daniel Sickles's Brigade at Pope's Creek September 24, 1861. He was apprehended by order of the Provost Marshal of Washington and brought to the city about 10 days later under orders of Colonel R. B. Marcy. The days between his arrest at home and delivery to the prison in the Capital were spent mostly out of doors in the vicinity of one of Hooker's encampments on the Potomac. This period of captivity was very uncomfortable and most aggravating to Jones as he described these days later to his captors in Washington. He was taken to the Capital by way of Piscataway, Prince George's County. The prisoner was seen here by some of the locals while on the way through to the city. [15]

After about three weeks in Mr. Seward's prison at 13th Street and Pennsylvania Avenue Jones wrote a letter to the Secretary asking for some clarification of his case. Apparently he had seen no written, clearly spelled-out charges. The letter of October 25, 1861 to William H. Seward follows in its entirety because it gives special insight into the character of Jones and his life as

---

[14] *RG 59*, November 27, 1861, Report L. C. Baker to W. H. Seward, Investigation of Affairs in Southern Maryland.

[15] *W.R.O.R. II-II*, p. 864.

life as farmer and boatman on the Potomac. Jones's presentation of his case reveals an educated man who had little trouble putting his feelings and their related facts on paper. The words written to Seward flesh out the total make-up of Jones as an above average personality of his time, place and purpose. The letter was written while he was in a Department of State prison some distance from Old Capitol Prison and near the commercial center of Washington, D. C.:

> *"Corner 13th Street and Pennsylvania Avenue,*
> *Washington, D. C., October 25, 1861*
>
> *Hon. William H. Seward, Secretary of State.*
>
> *Dear Sir: I understand my case is before you for investigation. Not having a chance to defend my case I have thought it would be well to adopt the following, viz:*
>
> *The charges against me as I understand are as follows: That I have been engaged in ferrying persons to and from Virginia, carrying arms and ammunition to Virginia; that I kept a horse and carriage to convey persons to and from my place; that I had been to Virginia myself and when arrested I had a double-barrel gun loaded with buckshot, also a revolver with me. The revolver I have had since 1856 for self-protection; the double-barrel gun was left with me until the owner called for it. I knew nothing of its charge until it was discharged in my presence after my arrest. I understood that troops were scouring the counties and making arrests, and I as well as many others went to Virginia to escape arrest. Hearing the troops had left I returned home. I had never taken up arms and never intended so to do. The horse and carriage was left with me by a man by the name of McKenny who said he belonged in Washington. He brought with him in said carriage five runaway Negroes which belonged in Virginia to a man--I forget his name. Said Negroes were released from prison in Washington by order of, as McKenny said, General Mansfield or General Scott, which of the two I have forgotten. The horse and carriage were to remain with me until he [McKenny] returned from Virginia. If there were arms or ammunition carried to Virginia I have no knowledge of the same.*
>
> *I do not deny but I have carried some persons across the river to Virginia. I carried several families of mothers and children to Virginia whom they said had husbands living there. Some of them said they had passes but I did not see them. I carried men also. I never inquired their business. I had not carried any person to Virginia for some two or three weeks before*

I went there myself.

I further understand that I am charged with carrying horses across the river which is certainly as false a charge as any man ever was charged with. After losing two very valuable boats by the Government I repaired a small-bottom boat for my express use; that is to say fishing and crabbing. Persons frequently came to me after said boat was repaired to get me to carry them to Virginia. I positively refused upon the ground that I had her repaired for my own purpose and if I attempted it I would lose her. A great part of the work was done by free Negroes which is known to be a fact; and furthermore every man on the Potomac River that had a boat has been guilty of the same offense charged against me. Where there was a boat there was no use in saying "no" when men from a distance came and said they wanted to go to Virginia on important business and must go. I have known in several cases where they after being positively refused took the boat and crossed the river themselves. Apart from conveying men across the river there have been men arrested in my county and released that were engaged in recruiting and sending men to different points on the river to cross and giving other aid and comfort to the enemy.

I have said already more than I intended and more than necessary, but I being the only man from my county held under arrest as prisoner causes me to feel more mortified and to suffer more than I should have done if others had been made to suffer the same; but it is frequently the case that the innocent have to suffer for the offenses of the more guilty which is now my case. I am the only man of the county held as prisoner and consequently am paying the penalty of the whole county. What I did which seems to be treason to the Government I did for profit which was not the case of many in the county. They fed, transported and furnished money for the Southern cause. I was not able to do that and did not do it. All I did I got paid for which I do not think I ought to be blamed for. I have a large family to support, and being a poor man I thought that if I could make something by carrying a few persons across the river it would be no harm; but instead of profiting by the operation I have lost severely. I have lost two valuable boats, one fine horse, saddle and bridle and revolver that cost $20. That of all most valuable is my liberty and to be separated from a distressed wife and eight little children. Also I have lost much by being from home at this important season of the year when a crop was to be taken care of and a crop of wheat to be seeded, besides a great deal of work that should be done very soon *before winter sets in. I have no one at home to carry on such work and no one that knows anything about a great deal of work I intended to have done this fall.*

The above is a long and uninteresting letter but I am in hopes you will see my object and sympathize with my family if not myself; besides I have had much suffering since my arrest. I was kept in camp eleven days, having nothing to sleep on but the naked ground and nothing but a blanket I fortunately brought with me, and but little that I could eat the biggest part of the time. I have been from home now five weeks next Tuesday. It seems to be a long time. I am willing to make oath on the Holy Evangels of Almighty [God] that the above statement is the whole truth and nothing but the truth. I am further willing to take the oath of allegiance.

Yours, very respectfully,
THOS. A. JONES' [16]

Josiah Dent of Charles County, a highly respected local citizen, tried to help county people in Washington jails after arrests on somewhat vague charges having to do with suspicions of treason against the Federal establishment. Writing to Dent November 18, 1861, Jones described the distress of his family back on the bluff near Pope's Creek. Jones tells Dent about the approaching time of confinement for wife Jane, expecting the birth of their ninth child. And he points out that he must get home soon to make plans for next years' crops and that it soon will be too late to do so...also, he has not provided his family with winter clothing and shoes, and cannot do so until he is released from prison. He desperately pleads for intercession. [17]

During November and December pressure to gain the release of Charles Countians accused or simply suspected of subversive works came to nothing. Seward stood firm and cold against entreaties by Charles County Judge Peter W. Crain and Maryland Member of Congress Charles B. Calvert. [18]

A letter dated December 31, 1861 to Jones from his sister Ann Mary sketches a very pathetic and grim picture of conditions in the Pope's Creek home with a father now imprisoned well over three months. Ann Jones announced that Jane was confined the day after Christmas and gave birth to a fine boy the 26th. And a direct quote,

---

[16] *RG 59*, October 25, 1861, Letter Thomas A. Jones to William H. Seward.

[17] *W.R.O.R. II-II*, p. 869.

[18] *Ibid*, pp. 873-874.

> *"... the children say he is quite a Christmas gift, nine little children is quite a number for a mother to have the care of without the attention of a Father ... we are most anxiously praying for you to come home ... Jane tries to have things properly attended to... she has but little influence with the boys ... she has many trials but I hope her Heavenly Father will give her strength to bear them ..."* [19]

Shortly after the first day of the new year Jones and the others from Charles County were transferred from the 13th and Pennsylvania Avenue accommodations to the probably less comfortable quarters in the Old Capitol Prison across the street from the U. S. Capitol building. Early in January these men came very close to being released. Secretary of State Seward in an order dated January 9 instructed Brigadier General Andrew Porter to release all Charles County, Maryland prisoners confined in the Old Capitol Prison upon each taking the oath of allegiance to the Government of the United States, and the adherence to a few other stipulations. [20]

However, on January 11 Porter replied to Seward's note in these words:

> *"I would respectfully represent that the above-named prisoners* [Jones, the Dents, the Watkins and Harbin] *are of the most dangerous character, and have been actively engaged in furnishing information to the Rebels, and in transporting men to Virginia for the purpose of joining the Rebel army as also in the nightly transportation of contraband goods to the enemy, that the prisoners Dent and Jones were duly accredited agents of the Rebel Government for the purposes above set forth ..."* [21]

Behind Porter's rejection was a letter dated January 10 from Headquarters, Army of the Potomac, signed by the army adjutant J. Williams on behalf of Major General George B. McClellan, now commanding this army. It informed Porter that McClellan thought the Charles County men should not be released considering the character of the charges against them, as a matter of military necessity. [22]

On January 28, 1862, Jones wrote another letter, this time to Brigadier General Porter and/or E. L. Allen, begging them to help him. He now had been locked up for over four months and his letter reflects much desperation and a bit of hopelessness. There is no indication in the

---

[19] *RG 59*, Ann Mary Jones ltr 31 December 1861 to T. A. Jones.

[20] *W.R.O.R. II-II*, p. 875.

[21] *Ibid*, p. 876.

[22] *RG 59*, 10 January 1862 ltr J. Williams to Brig. Gen. A. Porter.

RG 59 materials that any of Jones's earlier letters were acknowledged or answered. Pleas on his behalf from rather prominent personages in positions of respect and power seem to have been equally ignored by Mr. Seward and the Provost Marshal of Washington. Had Jones known at this time about McClellan's request that the Charles County group not be released his mental condition may have reached tragic limits. The letter read,

> "... *a few lines to you to solicit your sympathy if not for myself, for the sake of a distressed wife and nine children, one of which is a stranger to me it having been born since my imprisonment ... give my case an impartial consideration and grant me an early release ... located as my place was there is not twenty men in my county or the adjoining counties that would not have done the same as myself.*" [And, Jones goes on], "*... I do not know who my accusers are, nor do I know what the charges are ... I should have had a trial.*" He is extremely bitter over the apparent fact that many others arrested by Seward men for similar transgressions and perceived treasonous acts served much briefer periods of incarceration in spite of having much less family responsibility at home." [23]

Jones remained a prisoner in custody of the State Department from September 24, 1861 until February 15 the next year. On this letter date he was taken from State Department charge and turned over to the War Department and remained in Old Capitol Prison until finally released on March 21, after signing the standard Government Oath of Allegiance, as did the Dents, Watkinses, and Harbin the same date. One cannot but feel that all parties connected with the oath signing did so with tongue-in-cheek. Available records for this period do not hint at what strange twists and turns of Washington bureaucratic minds led from the McClellan order to hold Jones and his southern Maryland companions almost forever to a decision to free them after taking the oath in which they all swore "*... to support, protect, and defend the Union and Constitution and the Government of the United States as established by the Constitution, against all enemies, whether foreign or domestic ...* " etc, etc, etc. We know now that Thomas Austin Jones's contributions and sacrifices on behalf of the Confederacy went on throughout the next three or more years. Today it seems impossible to believe that Washington, after its 1861-62 experiences with Jones, would have removed their eyes from him for a moment. But apparently this happened and Jones went on his sometimes not so merry way in support of a cause that never lived up to his expectations for it. [24]

The Army's refusal to comply with Seward's order to release Jones and comrades seems to have marked the beginning of the end of the Department of State's role in handling matters of

---

[23] *W.R.O.R. II-II*, pp. 878-879.

[24] *RG 59, Oath of Allegiance*, March 21, 1862, Thomas A. Jones.

I, _Thomas A. Jones, of Charles County, Maryland_ do solemnly _swear_ that I will support, protect, and defend the Union and Constitution, and the Government of the United States as established by that Constitution, against all enemies, whether domestic or foreign; and that I will bear true faith, allegiance, and loyalty to the same, any ordinance, resolution, or law of any State Convention or Legislature to the contrary notwithstanding; and, further, that I do this with a full determination, pledge, and purpose, without any mental reservation or evasion whatsoever; and, further, that I will well and faithfully perform all the duties which may be required of me by law: So help me God.

Tho. A. Jones

Sworn to and subscribed before me, this _21st_ day of _March_, 1862

John A. Dix
Maj. Gnl.

Thomas A. Jones Oath of Allegiance to the U.S. Government signed March 21, 1862 which ended Jones's first incarceration operative E. J. Allen at Headquarters, Provost Marshal General, Washington City, wrote to the new Secretary of War, Edwin M. Stanton, as follows:

> *"Sir, Thomas A. Jones is the officially accredited agent of the Rebel Government, for conveying information and material thereto. In his house and that of his confederate George Dent, Senior, at Allen's Fresh, Maryland were found the signal regulations then in use between them and the Rebels. Jones is a most dangerous man to be at large even for the shortest length of time."* [25]

Old Capitol Prison in Washington, D.C. on site of the present day Supreme Court of the U.S. Constructed about 1815 to serve as the Capitol Building during the repair of damage done to the Capitol building during the British 1814 occupation of the city.

---

[25] *W.R.O.R. II-II*, p. 880.

PART 3

UNDER THE FEDERAL PRESENCE

The Jones open-boat ferry transports plied the Potomac with little hindrance for several months after the war began. Most of 1861 the government was so concerned about building up Washington defenses that the down-Potomac rebel-held regions were given little thought. Communication links between the Confederacy and points north were forged, strengthened and used with little risk by pro-South people and factions. Maryland itself during the first several months of the war remained occupied with matters hotly contested that related directly to the State's own status in the rebellion. In fact, everything appeared quite rosy to southern Maryland confederates after the amazing, disastrous rout of the Yankees at Bull Run.

Fortress Washington was not long in recognizing the weakness or non-existence of approaches to the capital along the Potomac. Hadn't the British Navy come within spitting distance of the White House less than half a century earlier along this same water highway and in spite of some formidable fortifications? With this awareness the government in early fall of 1861 ordered Brigadier General Joseph Hooker to begin moving his yet undermanned new division from Bladensburg campsites southward along the river. He encamped south of the mouth of the Mattawoman Creek, along the shore near the Chicamuxen neighborhood of Charles County, about 25 miles directly south of the District of Columbia. With three to five thousand troops permanently settled in here, and with better protection and bases for cavalry patrolling, things began to fall apart for those used to somewhat carefree use of the Potomac highway to send and receive supplies, news and men to and from Southern states.

A hint about the approaching reality of extensive Federal Government presence in the lower Potomac River counties took place early in June when about 100 Union troops landed at Chapel Point from the steamer *Mount Vernon*. They moved quickly to "Rich Hill," about five miles inland, and searched the house of militia Captain Samuel Cox, demanding he give up any State arms he held there. They found none and so immediately returned to Chapel Point for the trip back up river. [26]

Also in June local pro-Confederacy leaders took part in a Southern Rights Convention in the 6th (local) Congressional District. General Walter Mitchell's name was submitted as the man to represent the district at the State Convention. Prominent local lawyer Frederick Stone and Dr. Stoughton W. Dent were also at the meeting. [27]

And another pertinent note in the June 13 *Times* pointing toward the armed confrontation coming . . . a Confederate (artillery) battery had just been discovered near Mathias Point . . .nearly

---

[26] *Abstracts*, Volume II, p. 136.

[27] *Ibid.*

Early winter scene in Charles County camp of 8th New Jersey Volunteers near mouth of Mattawoman Creek - General Hooker's Division. By Alfred R. Waud, March 1862

opposite the mouth of the Port Tobacco River.

Early in September word reached the provost marshal from Pope's Creek that many people were being ferried across to Persimmon Point in Virginia from the Maryland shore. About mid-August some Union sailors from the gunboat *Pawnee* had been shot here while going ashore to check on a beached small boat they thought ready to leave for the opposite shore. Friends of the Union visiting relatives near Pope's Creek advised Washington that Yankee troops had visited that neighborhood about August 25 and confiscated "Tom Jones's boat" . . . also that he had gotten another one a few days later and continued to take people across the river night and day. Now, Jones began keeping his boat on the Virginia side away from the growing threat of Yankee confiscation. We must marvel at the nerve and total courage of Jones. Simply putting oared boats across the Potomac in all kinds of weather was usually a risky business. [28]

The first indication that the Union was through playing games along the river came with the September 23, 1861 arrest of George F. Harbin in Washington. This Harbin was a brother of Mrs. Thomas A. Jones. A known Southern sympathizer, Harbin was charged with writing denunciations of the Federal Government and promoting Southern victory. Harbin and several others with roots in Charles County were arrested in Washington about this same time and ultimately ended up in Old Capitol Prison. Members of the Watkins family also were arrested in Georgetown and charged with spying and helping men cross the river to join the Confederate Army. They too ended up in Old Capitol Prison. (This building was constructed about 1818 on the site of the present Supreme Court building.) [29]

Then, the same blow fell on boater-farmer Jones. He was arrested September 24 by troops from Brigadier General Daniel Sickles's Brigade of Hooker's Division, which were just settling into new campsites along the Potomac. They were in the Budd's Creek area overlooking the river toward present-day Quantico about a mile and a half away. Quite likely Jones was picked up at his home near Pope's Creek, about nine miles by water from the village of Port Tobacco, then the county seat for Charles. He was charged by the Provost Marshal at Washington with furnishing information to the rebels, carrying Marylanders over the river to join Confederate armed forces, and nightly transportation of contraband goods to the people of Virginia. [30]

In a continuing crack-down on those supporting the South, Jones's near neighbor George Dent, Senior, was arrested about mid-November while in Virginia directly across the river from his home at Pope's Creek. He was accused of transporting men, arms and ammunition to rebel

---

[28] *W.R.O.R. II-II*, p. 862.

[29] *Ibid*, pp. 857, 858.

[30] *Ibid*, p. 858.

Action shot of U.S. Navy's first aircraft carrier anchored in mouth of Mattawoman Creek just off Potomac River in early fall of 1861. This balloon launch platform was converted in the Washington Navy Yard from the former riverboat excursion steamer, *George Washington Parke Custis*. The balloon observers were sketching the locations of Confederate artillery batteries on the opposite Virginia shore. Courtesy National Archives.

contacts in Virginia while an authorized agent of the Confederate Army. [31]

By the end of the year Hooker's Division had received its full complement of troops...three brigades: First Brigade . . . Hooker's own command, consisted of 1st Massachusetts Vols., 11th Massachusetts Vols., 2nd New Hampshire Vols., and 26th Pennsylvania Vols.; Second Brigade . . . commanded by Brigadier General Daniel E. Sickles, consisted of the 70th, 71st, 72nd, 73rd and 74th New York Vols., and the division's Third Brigade of New Jersey Vols., composed of the 5th, 6th, 7th and 8th Regiments, all under command of Colonel Samuel H. Starr. Supporting Hooker's Division were: Battery D, 1st New York Light Artillery, and most of the New York State 4th and 6th Light Batteries. Battery H., First U.S. Artillery (Regulars) also was attached to the Division in its final posture along the Maryland Potomac shore. [32]

The Second New Hampshire Regiment of Hooker's Brigade was the first unit of that size assigned to the new division at Bladensburg. Troops of this regiment explored the Charles County countryside far inland from the river. About October 28, with supporting light field artillery of Doubleday's Battery (Battery D, 1st New York Light Artillery) they were detached from the main division camp and moved to Hill-Top, about five miles west of Port Tobacco and about half an hour's gallop from Jones's Pope's Creek home base. At this time Hooker's assigned scout support troops, the 3rd Indiana Cavalry, made the strengthened Federal military presence felt from the mouth of the Mattawoman all the way south to Point Lookout. Company E of the 3rd was ordered to headquarter near Port Tobacco. The 3rd Cavalry Yankees patrolled many of the main roadways, and even inlet waterways, constantly, until recalled by Hooker to its Budd's Ferry base in March. They remained there until the 24th, when they returned to Washington.

The 3rd Indiana Cavalry never joined the great McClellan exercise on the Peninsula where Hooker's men headed from their southern Maryland bivouacs in March and April, 1862. [33]

The Union troop concentration and disposition in southern Maryland is detailed very well in *Fighting Joe Hooker*, written by Walter H. Hebert and published in 1944. This work indicates that most of the Yankee combat experience along the Potomac in Charles County occurred in camp . . .between regiments and their commanding officers. Hooker himself seems to have been a leading instigator of bickering and breakdown of command functions and respect for commanding

---

[31] *Ibid*, p. 859.

[32] *The Equestrian Statue of Major General Joseph Hooker*, Erected and Dedicated by the Commonwealth of Massachusetts. Printed by Order of the Governor and Council, (Boston: Wright and Potter Printing Co., 1903), p. 198.

[33] Descriptions of Hooker's Division adventures in southern Maryland, October 1861 through March 1862 are based primarily on accounts in regimental histories of *Third Indiana Cavalry*, W.N.Pickerill, Indianapolis: 1906; *Second New Hampshire Vols.*, Martin A. Haynes, Manchester: 1865; *A History of the Second Regiment, New Hampshire Volunteer Infantry in the War of the Rebellion*, (Lakeport, N.H.: 1896) (also by Haynes); *History of the Third Regiment, Excelsior Brigade: 72nd New York Volunteer Infantry, 1861-1865*, (Jamestown, N.Y.: 1902); *A Narrative of the Formation and Services of the Eleventh Massachusetts Volunteers from April 15, 1861 to July 14, 1865*, (Boston: Gustavus B. Hutchinson, 1893).

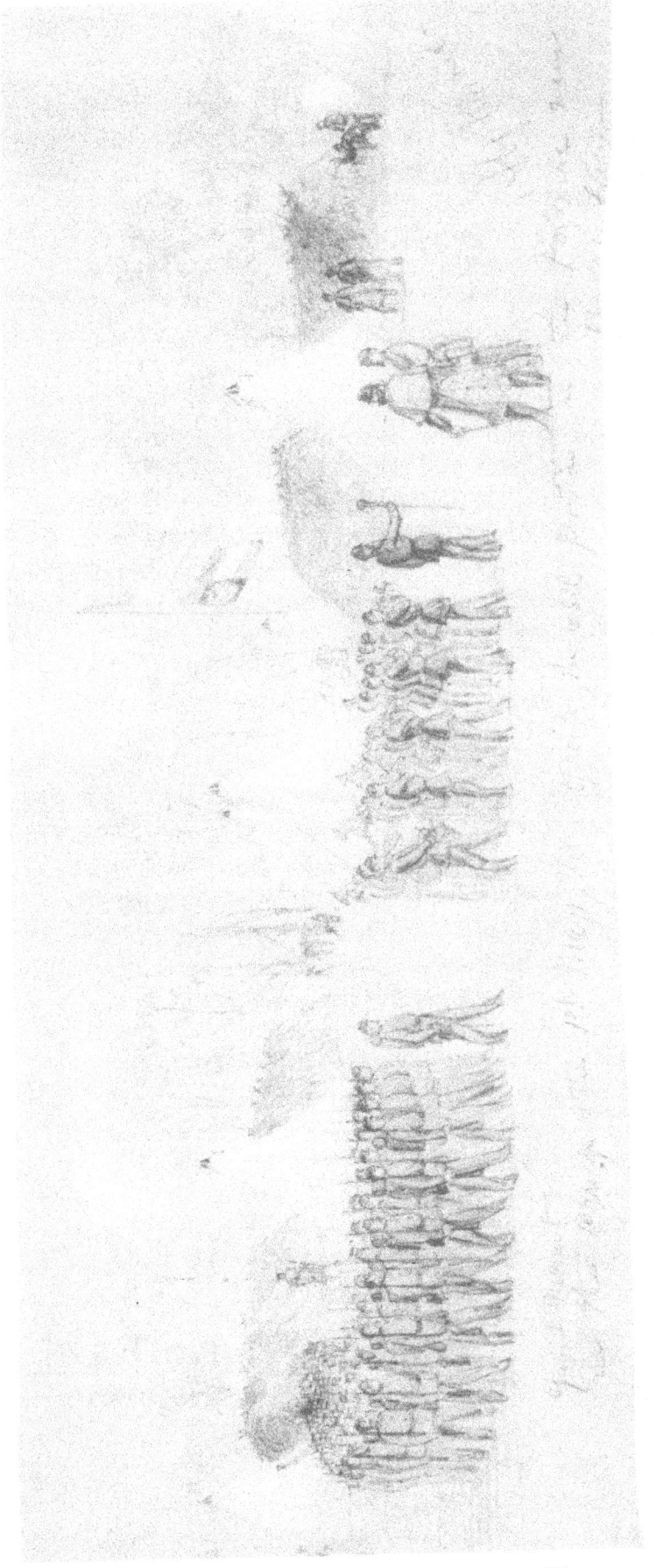

Scene sketched sometime from November through February 1861-1862. This shows the guard mount in the camp of the 1st Massachusetts Volunteers who were camped near Budd's Ferry on the Potomac River in Charles County about 25 miles south of Washington, D.C. Sketch by Civil War artist Alfred R. Waud. Courtesy Library of Congress.

officers. And, heavy consumption of alcoholic beverages took a heavy toll among Union troops in southern Maryland. Hooker's headquarters troops were established along the Potomac in semi-permanent bivouacs near the end of October, about six miles northeast of Budd's Ferry. Troops were initially posted in all strategically vital locations in Charles County, but primarily in the western half. Large warehouses and a wharf were built quickly at Rum Point on Mattawoman Creek. Another warehouse soon went up at Liverpool Point, just south of the Mattawoman. And, a good sign of long-term occupation was seen when Northerners established telegraphic communication with Washington headquarters of General George B. McClellan, General Joseph Hooker's commander.

While Jones spent most of the 1861-62 winter under cover in Federal prison, Union troops along his river found just how unbearable a bad winter could be in southern Maryland. The weather became sharp and increasingly frosty early. On January 6 the Potomac froze shore to shore, cutting the critical supply highway to Washington. Then, in a few days rising temperatures made roads impassable...local roads, never really adequate, turned to trenches of bottomless mud. In fact, many Yankees probably felt as much like captives as overbearing occupiers in their seldom comfortable quarters along the Potomac. Similar weather and road conditions in late 1864 quite likely led to a desperate decision to murder rather than kidnap Abraham Lincoln. And it probably is true that lowly, ugly, useless mud shoved Thomas A. Jones smack into the middle of one of America's most infamous tragedies.

At about the time several Charles County southern sympathizers were taking the Federal Oath of Allegiance in Washington and about to head back home, a daughter of one of them wrote a revealing letter to a friend across the river. Miss Sophie Ashton Dent, daughter of George Dent, Senior, of Prospect Hill at Pope's Creek, wrote to Miss Stuart directly across the Potomac in King George County:

> *"I am very sorry that I have not been able to send the gray flannel which you requested me to purchase for you, but I was not so situated at the time as to comply with your request but when the next boat comes over I will send the newspapers, the tea and flannels, or anything else I can procure which will be useful the other side of the water. I expect my father home before long. If however he takes forty oaths of allegiance he cannot prevent my shipping what I please, that is without he is much sharper than I have ever known him to be before. Therefore, know that it will give me the greatest pleasure to serve you at any time."*

No doubt Miss Sophie was disappointed beyond measure when her spirited note fell into the hands of a trooper of the 3rd Indiana Cavalry. If nothing else, this note reflects the almost total failure of Hooker's force to intimidate his Maryland neighbors along the shore. [34]

The letter below touched conditions and special features of Jones' final days as a Government prisoner:

"Old Capitol Prison,
Washington, D. C.,
February 28, 1862.

Hon. P. W. Crain, [Charles County Circuit Court Judge]

Dear Sir: In The Star of yesterday I find an order issued by order of the president appointing two commissioners to examine our cases. One of the commissioners I find to be Maj. Gen. John A. Dix who I understand is a gentleman of liberal feelings. You will please see him and try and make a favorable impression in my behalf. I received a letter from home dated 13th instant informing me of the illness of my wife. I received a private message yesterday that she was still ill. There are several reasons why my case should be acted on favorably. First, the condition of my family; second, my services are wanted on the farm as it is now working time; third, that the Government has between $400 and $500 worth of my property; fourth, I have been a prisoner for upward of five months and the most of the time I have been confined in a room where there has been some one sick all the time, and some of the cases have been of a malignant form. The last statement can be proven by the medical attendant and supervisor of the prison which I have been confined in.

Please give the above your especial attention as it may be the last chance for some time. I am in hopes to be able to give you satisfaction for all the trouble you have taken for me.

Yours, respectfully,

Thos. A. Jones" [35]

The following order ended Jones' imprisonment.

---

[34] Walter H. Hebert, *Fighting Joe Hooker*, Chapter IV, "In Lower Maryland," (New York: Bobbs-Merrill, 1944).

[35] *RG 59.*

> W. P. Wood, Esq.,
> Supt. of the Old Capitol Military Prison, Washington, D. C.
>
> *Sir: You will please discharge * * * George Dent, Sr., George Dent, Jr., Thomas A. Jones and George S. Watkins, prisoners confined in the Old Capitol Military Prison, on their taking the oath of allegiance to the Government of the United States.*
>
> *Very respectfully, yours,*
> *John A. Dix,*
> *Edwards Pierrepont*
> *Commissioners"* [36]

Washington Federal authorities sometimes extracted information from Black prisoners who usually were more voluntary with details about what they had seen while behind Confederate lines and among Confederate sympathizers. One such person was William Hill from the Mathias Point section of the Virginia shore. He furnished the provost marshal with some interesting details about the ferrying operations conducted by Jones and Dent from Pope's Creek. Hill was a former slave of James Grimes. At this time Jones himself owned slaves who often were pressed into service for their master in the Potomac transport business. Evidently, Jones and Grimes ferried only people, probably since the craft needed for horses and wagons were too large and slow for speedy passage necessary to elude Union Navy river patrols. [37]

Late in October Reverend Wilmer would have had an opportunity to communicate face to face with Federal authority. A regiment of Sickle's New York troops visited Port Tobacco and camped at "Mulberry Grove," the home of Dr. Robert Fergusson. They had tarried an hour or so in Port Tobacco and "*. . . conducted themselves very properly.*" The *Times* editor stated in the October 31 issue that this group would join a brigade along the Potomac River numbering 10-12,000 men, stationed from Mattawoman Creek to Budd's Ferry. Union artillery were reported at Hill-Top, about eight miles west of Port Tobacco a few days earlier. [38]

The shore-to-shore communication machinery was extremely critical for the safe and effective performance of the ferry system. Jones described some of this in his Columbian Exposition work of 1893. Prisoners reporting to the provost marshal in the capital added some detail about the system as it worked on either shore. Because Jones himself never was apprehended by Federal river craft during a crossing we must believe the signals worked very well

---

[36] *Ibid.*

[37] *W.R.O.R. II-II*, p. 862.

[38] *Abstracts*, Volume II, pp. 140, 141.

throughout the war. Crossings seem to have been mostly in the Maryland to Virginia direction, so signs on the western side were the most common and critical . . . whether safe to go or not to go was the big thing. And with visibility normally splendid for miles in either direction up and down river it usually would have been possible to reverse course and return to shore should unfriendly vessels be spotted approaching from either direction. Night time and fog-screened crossings were another matter . . . but discretion, sharp lookouts, and muffled oars gave the ferrymen a usually sufficient safety cushion. Passengers from Pope's Creek usually found horse-drawn vehicles waiting at Grimes's to take them wherever headed in Virginia. A good rider with a fast horse should have been able to reach Richmond in three to four hours after landing. On clear nights a lighted window at either Grimes's or Watkins's was the sign that somebody needed to cross.

It is difficult to determine whether or not Jones enjoyed official Confederate government authority for his work during 1861. Nor do we know for sure how much, if any, compensation he received from Richmond. Probably the passengers themselves paid the service charge and the cash income doubtless was the main motivation in conducting ferry services. Jones's letter of October 1861 (while in Department of State custody) to Seward indicates that the Pope's Creek ferryman functioned as a trans-Potomac transporter much as he had for some years...as many other small-time entrepreneurs had been doing up and down the river at places where relatively rare and expensive regular ferry service did not exist. At many points along the river service such as the Jones' ferry operated on an as needed basis. For three centuries ferries had come and gone. A short distance south of Pope's Creek at present day Morgantown the Laidler-Hooe ferry remained in service more or less constantly from late 17th century. It ended with the 1940 completion of the Harry W. Nice bridge that joined Maryland and Virginia for the first time south of old Long Bridge (14th Street bridge--Rochambeau Memorial) in the nation's capital.

In 1890, Jones, then living in southeast Washington, D.C., applied for membership in the Society of the Army and Navy of the Confederate States in the State of Maryland. The membership application form shows that he officially entered the Confederate service in April 1863. He wrote that his position title was Chief Agent of the Secret Service (of the Confederacy) in Maryland . . . this position title served as his rank. His war-time commanding officer, Colonel William Norris, was Chief of the Signal Bureau (CSA). Colonel Norris himself endorsed Jones's application with a most revealing and commendatory note. He wrote:

> *"I certify, on honor, that I know of my own personal knowledge, that the above applicant served honorably in the Army or Navy of the Confederate States as Chief Agent of the Secret Service Bureau in Maryland where his unpaid services were of incalculable value to the Confederate States in keeping open the most thoroughly reliable path of communication through the Yankee lines for 2 ½ years . . .during which time the man lived under Yankee fire. ."*
> [39]

---

[39] *C.S.A. Maryland Records*, Maryland Historical Society, Baltimore. Thomas A. Jones Application for Membership in the Society of the Army and Navy of the Confederate States in the State of Maryland.

What more might any authoritative person have said to endorse the immeasurable value of Thomas A. Jones to the Confederate cause? We must add to this crediting of devotion and courage the equally important work of the previous two years, apparently without any meaningful official attachment to the Richmond government. One may hope that the earlier years of service were acknowledged in substantial way by Confederate States of America...but this is to be doubted. By way of limited recognition of long, faithful service rendered in a hopeless cause Jones was elected to membership December 16, 1890 in the above-named Society.

A substantial, extremely regrettable loss for researchers of southern Maryland Civil War history lies in the disappearance of even archival copies of over 150 issues of the *Port Tobacco Times and Charles County Advertiser* published in the war years. The sole remaining accumulation of *Port Tobacco Times* issues in Charles County was donated by the late Louis M. Hyde of La Plata. about 1940, to the Enoch Pratt Free Library in Baltimore. Here the surviving issues lay until about 1974. After micro-filming Enoch Pratt destroyed all original copies in their possession apparently without informing anyone in Charles County. A few original copies of the *Times* remain in Charles County and in the Hall of Records in Annapolis.

Thanks to the editor of a neighborly publisher in adjacent Prince George's County, we can account for what happened to one issue of a fall 1861 edition. *The Planters' Advocate and Southern Maryland Advertiser*, published in Upper Marlboro, Maryland, October 9, 1861, printed an item titled: "Affairs in Charles County." It solves the mystery of what happened to missing Nr. 22 of Vol. XVIII, dated September 26, 1861.

> *"We copy the following item from the Port Tobacco Times of Thursday last* [issue of October 3]:
>
> *The day after our last issue* [wrote *Times* editor Elijah Wells] *a request was made of us for the use of our press and type, to issue a paper for the benefit of the soldiers encamped near this place. We urged no special objection, and printers from the regiment at once set to work for the purpose above named.*
> [A copy of this specialty item may yet exist in some New York State family archives or library, since the title suggests strongly the issue was produced by soldiers of Sickles's New York Excelsior Brigade.]
>
> *The rumors abroad that our office was forcibly taken possession of, and the press and type carried off, is incorrect.*
>
> *We learn that our friends in the county were for a while quite moved at our arrest by the military, and we embrace the first opportunity to return our sincerest thanks for the sympathy exhibited. We are, however, back again at our post, and though it may be thought that we are somewhat perplexed at the 'firey* [sic]

opportunity to return our sincerest thanks for the sympathy exhibited. We are, however, back again at our post, and though it may be thought that we are somewhat perplexed at the 'firey [sic] *ordeal,' we assure the readers of the Times that we came out 'without even a hair singed or the smell of fire about us.' Conscious of no wrong on our part we 'faced the music,' and are free to say that the officers clothed with authority to investigate our case acted with promptness, firmness and with an eye to justice, hence our speedy release. With this short explanation, we make our best bow to the readers of the Times, and resume our duty as caterer of items for the public."*

Jones's life from marriage until outbreak of the war is not well documented for somebody who risked and accomplished so much that was extremely demanding over a long period of time. Volume II of the *Abstracts from the Port Tobacco Times, 1855-1869* reveals more about the man than anything else available to scholars and writers. Tom Jones was farmer, boatman, fisherman, government worker, and a devoted Democrat in local politics. That he enjoyed some notable favor among the "Who's Who" in local politics is reflected in his role as tax collector, indicated in the 1850 U. S. Census, and in the composition of Charles County's grand and petit juries. In 19th century southern Maryland society jury service was seen as a desirable assignment because the compensation was good and in cash. And cash money in 19th century southern Maryland was a rare and wonderful thing. Further, jury duty in Port Tobacco often meant a stimulating, recreational break in the lives of those serving . . . only men, of course, and until about 1880, never other than White men. Jurors some distance from home during court sessions (say 10 miles or more) spent evenings and nights, all at public expense, comfortably with congenial companions from distant neighborhoods seldom seen between spells of jury duty. Life in the 19th century here for 90% of the population was hard, often cruel, and generally uncertain and with always the shortage of cash. And so, like many others of the yeomanry here, heads of households did many things to keep body, soul and kin afloat. And Thomas A. Jones did no differently. You could be quite sure that when Jones's "near kin" Samuel Cox, Senior was appointed jury foreman Jones would sit on that jury. (Apparently, the terms "foster brother" and "near kin" respecting the Cox-Jones kinship connection derives only from the fact that Cox had been suckled as an infant by Tom's mother for some time because Mrs. Cox could not provide such service).

Charles County land records show that Jones was part owner of about 400 acres at Pope's Creek when the war started. He and two others (Matthews and Weems) in the partnership lost the property in 1863. By date of November 2, 1863 the land was sold by Sheriff George A. Huntt to Mary King Somerville for $7,100. Records of the foreclosure indicate that the package had amounted to nearly 500 acres purchased by the group in the early 1850's. The June 24, 1858 issue of *The Port Tobacco Times* announced that Thomas A. Jones was *". . . selling land that was part of his farm known as 'Pope's Creek' 220 acres, on Potomac River nine miles from Port Tobacco, 2 ½ miles from Allen's Fresh. Steamboat landing adjoins land--two boats ply weekly to Baltimore and District."* By any contemporary measure the entire original package must have been judged a prime and valuable piece of land.

Then, as the days leading to Fort Sumter rolled by, Thomas A. Jones participated in county political activity leading to a convention in Annapolis or Baltimore that would deal with the course of action Maryland might take regarding possible secession. While Jones took part in political anti-Lincoln affairs the *Times* openly announced formation and training of county militia units in anticipation of conflict with the Federal Government: Mounted Volunteers of Charles County meeting at Port Tobacco, the Smallwood Riflemen, offered a resolution that

> *"...we hereby pledge our services to the State of Maryland for the protection of her honor and fair name under the Constitution of the United States, as given to us by our forefathers in its original purity..."*

And Bryantown Minute Men (foot rifle company) and the Nanjemoy Rifle Company were meeting to elect officers. Jones himself served as 4th lieutenant in the Allen's Fresh Rifle Company. [40]

In a bit of "side-bar" levity, the January 24, 1861 *Times* described something about a one-man, one-horse show in front of the court house. Militia captain Cox's horse, Grey Medock, followed trainer Emmons into the court house and into one of the second floor jury rooms and safely down again. They sure knew all about real "horsing around" in old Port Tobacco. [41]

In continuing his experience in local politics, Jones must have been wide awake and totally aware of where the tempers and politicians of the times were steering the Nation...by early 1861 already rendered into two parts, de facto. So he entered without reservation into supporting the Southern cause with all his heart, mind and earthly resources. He could not have dreamed just how disastrous and costly this would be to his family and their way of life.

As southern Maryland felt itself sliding toward certain and bitter conflict with a government long respected and revered few here raised a voice of caution and dismay...at that time only the very courageous and self-confident would have dared to do so. Yet, one voice from a respected man of substance was heard, thanks to publication in the October 31, 1861 issue of the *Times*, of the words of Mr. Thomas A. Millar of "Holly Springs," in the far western reaches of Nanjemoy, close to Riverside. He said, while agreeing to run for the Legislature,

> *"... A Southerner by birth, I love her people, even in their madness and folly; and hope to see the day when her erring sons, with penitent feelings, will return to the old homestead and renew their allegiance to that proud flag that will and must be maintained, for it is the symbol of liberty."*

---

[40] *Abstracts*, Volume 2, 1855-1869.

[41] *Ibid.*

On December 9, 1890 Jones wrote a letter to an officer of the Society of the Army and Navy of the Confederate States in the State of Maryland amplifying the details of his CSA service described on the earlier membership application form. Writing now from 719 12th Street, SE., Washington, D. C., Jones stated,

> *"After seeing you sometime since in sufferance to becoming a member of your Society, I sent an application for the same but perhaps I did not state my claim sufficiently strong to admit me as a member. In addition to what I stated in the application I further state that I contracted with Col. William Norris, Chief of the Signal Service of the Confederacy, to act as Chief Agent for Maryland to forward all dispatches and other papers connected with the Confederacy, and to furnish said government with files of northern papers which were supplied promptly with but little interruption, receiving said papers the next morning after their publication. Also it was part of my duty to aid all Confederate scouts and agents to and from Richmond who came with proper passes from the proper authority in Richmond, all of which Col. William Norris can vouch for.*
> *Very respectfully yours,*
> *(s) T. A. Jones"*

Thomas A. Jones about 55 years old.
Courtesy Carolyn Jones Straeter.

The above details of Jones's wartime CSA service give form, substance and legitimacy to his role as a major Confederate agent in Maryland during nearly all of the war years. It seems clear now that his personal security was at risk most of the time and that he most certainly did, as Norris wrote in his endorsement, "... *the man lived under the Yankee guns during his years of undercover work for the Confederate States of America.*" Not a day passed during which Jones might have been arrested, receiving the standard treatment given secret agents by the Federal Government during the Civil War years. It is to be wondered what Jones's innermost thoughts were during the Washington trial of the Lincoln assassination conspirators. But his friends and neighbors never deserted him . . . nor did his great spirit and boundless courage. And he must have guessed about mid-1864 that all was lost as General Ulysses S. Grant's Union Armies forged an ever greater and more invincible wall of military might between the Potomac River and Richmond.

Since Jones was responsible for helping legitimate Confederate scouts working in Maryland it seems he would have known quite a bit about the very small group of Mosby's Rangers who stealthily crossed the river in the fall of 1864 . . . on an errand of mischief, primarily. Led by Prince George's County native Lieutenant Walter (Wat) Bowie. This raid sometimes is described as "the Mosby invasion of Maryland." That seems a rather far-fetched appellation. Bowie commanded a company of Mosby's Rangers and was given permission by his commander to take twenty five men and try to do some scouting in Maryland. Bowie and two of his company crossed first for a brief reconnaissance. Then Bowie sent word to the Virginia shore that all of his group but eight were to return to Mosby in Fauquier County because the larger group would result in too hazardous an effort. The troopers' horses were left behind and Bowie planned to steal some Federal replacements from a small group of Union cavalry in Port Tobacco. This intrepid group of eight, armed only with revolvers and in full Confederate dress and with no supplies, hoped to reach Annapolis and kidnap the governor, or at least rob a bank, and then go around Washington and re-enter Virginia above the Great Falls of the Potomac.

While in Port Tobacco about to dispossess some Federal troopers of mounts Bowie and the seven others in gray reached the friendly comfort of the Brawner House and relaxed among sympathizers in the basement bar . . . waiting for darkness and quiet of the night. The Yankee camp was just outside the village, but a few members of the provost guard were lounging in the court house about 300 feet from the Brawner bar. In a 1912 book titled *On Hazardous Service . . . Scouts and Spies of the North and South,* by William Gilmore Beymer, there is an intriguing, lively, somewhat humorous account of how the Bowie raiders handled the Union cavalry troopers in Port Tobacco. In fact, the story may well have been told to the author by one of the ex-Rebels who participated. The whole thing reflects realism a bit beyond only an author's pure imagination. So, for color and a bit of "what things were really like," follows some direct quoting from Beymer:

> *"It was in the superfluous court-house that the provost guard was stationed . . . They [Bowie's men] strolled over to the court-house, and in the dark Bowie and Wiltshire [a Bowie man] throttled the single guard;*

> *then they all tiptoed into the wide, empty hall. In the doorway Vest [a Bowie rebel] was stationed. 'Shoot the first man out, Charlie,' Bowie said. There could be no retreat by Confederates, no escape for Federals-- Charlie Vest was a certainty when it came to shooting the first man. Just inside the hall [likely the courtroom] there were two doors, one on either side [leading to the Register of Wills' office to the right and Clerk of the Court's office to the left] and both rooms might be filled [with Yankees]. Walter Bowie and Wiltshire, lighting a match, flung open the door on the right-hand side and went in; the floor was filled with men...twenty of them. 'If any one fires a shot, we'll murder you all,' Bowie yelled. The rest of his men rushed in; the match went out, and in the dark and confusion it seemed that the entire Confederate army was charging into the room. Outside, Charlie Vest was waiting to shoot the first man. But none came out; the Federals were each giving a parole not to leave the room or give information till morning, 'when we will be safe across the river,' Bowie shrewdly said. The Union men kept their parole, though they had to listen to the raiders riding off on their horses.* [42]

Wat Bowie's thrust into Maryland ended tragically for him. Plans to kidnap the governor and take him to Richmond aborted. The idea of robbing a bank as next best thing also came to naught. The little, but spirited and dauntless, group, did however in desperation because of hunger rob a country store near Sandy Spring, Maryland (about fifteen miles directly north of the District of Columbia). This ill-conceived action in pro-Union countryside triggered a local response of unsympathetic protest that led to the killing of Wat Bowie and the dispersal of his Rebel force. Thus, ignominiously ended "The Mosby Invasion of Maryland." Before enlisting in the Confederate service Bowie had been comfortably and promisingly situated as a country attorney in Upper Marlboro, county seat of Prince George's County, a scion of the long distinguished and wealthy Bowies of Maryland. This was the family that played a major role in establishing the long promised Baltimore and Potomac Railroad, spawned before the Civil War but not completed until 1872.

A local description of the Wat Bowie incursion did not appear in print until carried by the February 12, 1897 *Port Tobacco Times*. A letter to the editor (signed Anne Tiquity) proposed several local history stories for the paper to publish, suggesting a continuing effort at writing up the "ancient history" of historic old Port Tobacco. For example, it was suggested the paper might

---

[42] William Gilmore Beymer, *On Hazardous Service, Scouts and Spies of the North and South*, Harper & Brothers Publishers, (New York: 1912), pp. 134-145.

Charles County Courthouse - Port Tobacco, Maryland
1819-1892

Wyatt Brown 1969 drawing of the reconstructed 1819 Charles County Courthouse which was destroyed by fire on the night of August 2/3 1892. This building was completed and dedicated in August 1973.

tell how "Watt" Bowie with a score or less of the "rough riders of the Confederacy" came over from Matthias Point and lay for one day in the woods near McConchie, fed by...Benton Barnes and Ned Smith, until night when they made a sudden dash on the old village, captured a company of Federal cavalry . . . drove them into the courthouse...picked the best of the Yankee horses and dashed on out of Port Tobacco. [43]

The next "Port Tobacco court house battle" occurred about twenty years after the Bowie assault. One ex-Judge D. R. Magruder (serving as an attorney) was severely beaten about the head by local attorney L. Allison Wilmer in a disgraceful display of unseemly juristic manners seldom seen in a courtroom. Wilmer accused Magruder of lying in court in the State vs Carpenter trial...one especially charged with emotion and tinged both with serious moral and political factors. Magruder was knocked to the floor and the sitting judge came down from the bench and restored order. Wilmer was removed from the courtroom by the sheriff. [44]

The life of Thomas A. Jones during the war years was one of personal danger and an ever downward spiraling of family fortunes. Jane Harbin Jones's death must have been terribly disheartening. For any parent left with so many children still in their minority life must have been crushing. Time taken away from farming responsibilities by the constant pressures of mail and dispatch handling, as well as management of the river ferry service, in themselves would have taxed the physical, spiritual and psychological limits of anyone. Constant fear of betrayal and discovery would have totally unnerved most men and their wives in a matter of months. Yet Jones survived and remained a most effective link in north-south communications and espionage operations. And so life went on for him, marred by loss of his fine home south of Pope's Creek and prime agricultural acres near the Potomac River bluffs. The family moved inland and a bit to the north of the historic large bluff-top home of the Dents, long called "Prospect Hill" which was destroyed by fire about 1917.

Dent family home "Prospect Hill" at Pope's Creek on bluff overlooking Potomac River. Photo taken about 1900 when property was owned by James Neale Hamilton. Library of Congress - Miller Collection.

[43] *Port Tobacco Times and Charles County Advertiser*, Volume LIII, Number 37, February 12, 1897.

[44] *Abstracts*, Volume IV, June 1, 1883, p. 184.

Tom Jones now lived directly across the Pope's Creek road from Thomas David Stone at "Ellenborough." He was a friend of Jones and staunch Confederacy supporter . . . also a gentleman farmer, slave owner and well-educated alumnus of Princeton University who was qualified to practice law but chose not to. He was born at Habre de Venture to William Briscoe and Caroline Brown Stone . . . exceedingly distinguished parentage.

An unforeseen, challenging, and exciting change in Jones's life began April 14, 1865 during a popular stage performance in Washington, D. C. attended by the President of the United States. In the very late evening of April 15 (or early 16) Jones got an urgent message from Samuel Cox, who lived at a plantation called "Rich Hill," about four miles southeast of Port Tobacco. Jones traveled quickly to the Cox home and was told by him that actor John Wilkes Booth and a friend had stopped there and asked for refuge. Cox then informed Jones that Booth had killed President Lincoln. Then he asked Thomas A. Jones to be totally responsible for the protection of Booth and his companion, David Herold. Jones, with no equivocation, promised to do what was asked of him . . . continuing business as usual . . . doing what he thought was in the best interest of the Confederacy, even though he knew full well now that it was all finished . . . he and the Confederacy had lost it all. The United States of America now appeared clearly to be victor after a long, terribly costly war between brothers, that set the course of history forever for a new nation begun in very uncertain circumstances in Philadelphia in 1776.

Thomas David Stone and his family lived at "Ellenborough" located on the Pope's Creek Road directly across from "Huckleberry."

PART 4

GATH AT CENTER STAGE

The story of Thomas A. Jones and his incredible sacrifices on behalf of the Confederate government is best told from now on in the writing of George Alfred Townsend. He wrote in the latter years of his professional life as "GATH." Townsend, Civil War correspondent, poet, historian, novelist, syndicated columnist, native of Delaware who, from the day of Lincoln's murder, began an almost life-long fascination with this subject, including the escape of Booth through southern Maryland.

Apparently, Townsend visited Charles County at least three times - 1870, 1877 and 1883. He visited Port Tobacco in February 1877 when he and former Civil War combat artist, Frank H. Taylor, covered a double execution in the village behind the courthouse. Then, both Taylor and Townsend (himself a competent sketcher) made some remarkably fine drawings depicting southern Maryland life, including the back wall of the Port Tobacco court house, street scenes in the village, and the Barnes-Compton house on the north side of the square next to the Brawner-St. Charles hotel.

Perhaps reflecting his circuit-rider father's Methodist liberalism, Townsend seems not to have backed the Confederacy. Most of his Civil War correspondent writing was a recording of Union Army experiences in Virginia, including the 1862 Peninsula Campaigns of McClellan's well-equipped, ill-led Yankee regiments heading for Richmond. Young GATH hit the popular American novel scene with his well received Eastern Shore-sited novel, *The Entailed Hat or Patty Cannon's Times*, which came off Harper Company presses in 1884. About 1870 GATH began research for a book based on the Lincoln assassination . . . John Wilkes Booth conspiracy. The writer's interest in this work took him through southern Maryland in a thorough, scholarly effort to locate people and places connected with the Booth flight, from Ford's Theater to the lonely little farm near Port Royal, Virginia. It was here the strange, hapless American stage near-great played out the unheroic final moments of his life stretched out on the rough boards of a farmhouse porch. GATH's southern Maryland discoveries found their way into his next novel, *Katy of Catoctin, or the Chain Breakers*, published by Appleton in 1886. [45]

Notable for historians who still often lose themselves in trivia more or less related to the Lincoln assassination subject, GATH reflected the feelings of many about the disturbing mystery of the disappearance of Booth and Herold for nine days after the brutal Ford's Theater balcony scene. Plagued by this incredible gap in the Booth escape episode Townsend visited Charles County in 1870 and there met Judge Frederick Stone (then a U. S. Congressman--40th and 41st

---

[45] George Alfred Townsend, *Rustics in Rebellion, A Yankee Reporter on the Road to Richmond, 1861-65*. Introduction by Lida Mayo, pp. XV-XVII, University of North Carolina Press, Chapel Hill: 1950.

Congresses) for the first time. The writer knew that Stone had been part of the defense team at the Lincoln conspiracy trial helping defend Herold and the accused persons from southern Maryland. Townsend wanted to find out what, if anything, the judge had been able to pry from the lips of Herold about his close association with Booth all the way from Washington, D. C. to Garrett's farm. [46]

GATH told Stone about 1883 that it was most important for him now to understand all there was to know about the great escape to and over the Potomac because he intended soon to begin some definitive writing about the entire assassination event. Stone told Townsend when they met in Charles County that year because certain people still lived who were involved in the escape story he could do or say nothing that would injure them in any way . . . that after their death if the future allowed him to speak freely he might do so. And eventually he did.

The Honorable Frederick Stone, Charles County jurist, member of the Maryland Court of Appeal, Member of the U. S. Congress and a most distinguished primary supporter of Confederate sympathizers in Charles County during the course of the Civil War.

Judge Frederick Stone's "Idaho" is located between Port Tobacco and La Plata. It was built about 1870 while the Judge served as a U.S. Congressman (1868-71) in the 40th and 41st Congresses. Here he lived with his second wife until his death in 1899.

---

[46] GATH obituary Special about Frederick Stone, written at Gapland, Md. shortly after Stone's death on October 7, 1899. Printed in major Maryland newspapers.

About mid-1883 GATH and Judge Stone met and spent a couple of days at his home, "Idaho," about half-way between old Port Tobacco and the new town of La Plata. Together, the columnist, now from New York, and the sedate, learned country judge of great pedigree traveled along part of the Booth escape paths and by-ways. However, because Thomas A. Jones still lived (in Baltimore) the Judge did not divulge the incredible secret. Nevertheless, Townsend found out a lot of the truth. Just how the writer explained things in his obituary written about the judge soon after his death --

> *". . . Judge Stone alone, of all the persons yet near that old event, was too considerate to reopen it. When we arrived however at Coxes store* [now Bel Alton] *younger persons, some of them of that very* [Cox] *family, volunteered what the judge had declined to give. The next day* [Sunday], *while his family worshiped at the old Port Tobacco Episcopal Church, I obtained from young men gathered at the courthouse* [in Port Tobacco] *the clue to a man then living who restored nearly a whole week of those missing days, not unimportant in the posthumous biography of a great man."* GATH closed the obituary with these words: *"Courthouse, witnesses, nearly everything, have now* [late 1899] *perished."* [47]

In a June 26, 1883 edition of the *Cincinnati Enquirer* a GATH article took up nearly half of page one. In it the author produced a fascinating column about a recent visit to southern Maryland. His reporting gives us one of the most scholarly, incisive accounts of life here in the turbulent years immediately after the War. He described riding the recently completed Baltimore and Potomac Railroad all the way to the Potomac River at Pope's Creek where it ended at a terminal built just a few feet from the water's edge. He wrote:

> *". . . Almost at this point John Wilkes Booth and his boyish companion, Harold [Herold] made their camp while waiting to cross the river into Virginia...About five miles from Pope's Creek to the northwest is an old, decayed country town called Port Tobacco where I went to church last Sunday. In that town and on the river below it lived a nest of contraband runners during the war, who took goods into the Confederacy, ferried over rebel agents and mercenaries, as well as persons who had some equitable excuse for breaking the blockade, such as lawyers and bankers. From Pope's Creek the great rebel secret line was established to Canada, connecting the Confederacy with the outer world in the most reliable way. Right at Pope's Creek lived two men conspicuous in this business, George Dent and Thomas A. Jones. They are still living.*

---

[47] Ibid.

"After the war, for some reason, he [Jones] was taken up by Senator Gorman's political machine and made a grain inspector in Baltimore, and afterward a policeman in that city, and next was given a position in the House of Corrections south of Baltimore. In the recent reformation of Maryland politics by Governor Hamilton, he has been turned out of office and is said to be collecting from his old neighbors material for the true narrative of Booth's adventures in Maryland after he left Dr. Mudd's house, about which very little is publicly known.

"Mr. George Dent lived on a high hill right above the river Potomac [north side of Pope's Creek], and his home was a place where the rebel night-hawks slept and watched their chance to get over the river. Some say that flash lights communicated from this house to the opposite shore of Matthias' Point. Dent lived in a small story and a half house with a front piazza which stood on a bluff. [In fact, the principal Dent house, which burned about 1917, was something of a very distinguished looking mansion constructed before the Revolutionary War and repaired after that war because of British naval bombardments]. He systematically ran the blockade during the whole War and if he had been a man of observation and intelligence equal to his adventures he could make quite an interesting book. In the neighborhood lived Dr. Dent [Stoughton W.], now an old man who also had a hand.

"The leading man in that country however and probably the chief spirit of the underground line, was Col. Samuel Cox, who died a few years ago, and whose will I read in the Port Tobacco Courthouse, and talked with his nephew and adopted son, who maintains his establishment and takes care of his widow, a lady who goes by the singular name Walter Ann Cox. [The Samuel Cox home, "Rich Hill," today in much reduced form, lies at the northern edge of the small Charles County community of Bel Alton on the Bel Alton-Newtown Road about four miles south of La Plata.] In Port Tobacco town, which now has just 200 inhabitants, lived [George] Atzerodt, who was executed on the gallows with Powell [Lewis Payne] and Harold and Mrs. Surratt. He was a Virginian, German, born in this country but speaking a dialect like the Pennsylvania Dutch, who came to Port Tobacco sometime before the War to follow his trade of carriage builder; but having very little discretion and being intemperate, he worked irregularly and as soon as the war began was drawn into the contraband business . . . he finally fell into Booth's clutches and ended his days on the gallows, after dying, as his neighbors told me, a *hundred deaths by fear.*

"... The man who probably helped Booth get out of Maryland was chiefly instrumental in building the railroad [Baltimore and Potomac] through the Western Shore. Mr. Samuel Cox was of an old family in these parts, a plain man, with a consumptive tendency, and this seemed to give him a desperate will and imperious spirit. To keep his lungs in good shape he had to drink more or less, and he was the driver of the region. He accumulated a small fortune, perhaps $30,000, but that is regarded as something considerable here. He was a raw-boned man, strong and active, with grayish hazel eyes, very smart and secretive, fierce, indomitable. At the beginning of the war he was made the captain of a militia company at Port Tobacco, the only one in his county. The lieutenant of the company was the brother-in-law of Dr. Samuel Mudd, one of the conspirators. One of Mr. Cox's nephews went into the Rebel army and was mortally wounded a short time before the crime of Booth, and he died only eight days after Mr. Lincoln's murder. Cox appears to have given all his energies to the help of the Confederacy, but was so cool and discreet that he was not found out. His house is a large and neat building, with a considerable extension on the ground floor, making another building, and between the two buildings runs a wide hall. The house is frame, filled in with brick [nogging] and the roofs of the two buildings are painted red. There are double yards to enter before getting to the house. After some conversation with the present inhabitants, he [Samuel Cox, Jr.] said to me, 'I have no objection to telling you, sir, that Booth and Harold came to this house about one o'clock on Saturday night [April 16]. They reached Dr. Mudd's house on Saturday morning [April 15], and stayed with him until dark and they were brought to our house by a Negro named Oswald Swann, [a Piscataway Indian] who is still alive. Oswald knew my father, and [that] he had a brother who was in the contraband trade on the Potomac River. I had been to Port Tobacco that Saturday afternoon, and there heard of Mr. Lincoln's death, and came home and told Pa. It was a bright moonlight night when we heard a knock on our door, and Pa put his head out of the window and asked who was there. A man requested him to come down, Pa asked the man who he was. He said he and another man, who was a little distance off in the road, wanted to come in. At that time the Negro man on the horse was outside of the outer gate, and we could not see him. Pa said he could not admit persons to the house unless he knew just who they were. They refused to tell him anything about themselves. The

parlay went on for half an hour and everybody in the house was up and heard it. The colored man afterward swore that these persons [the two strangers] stayed four hours in our house, and were entertained with champagne and food. At the time Harold told Pa that his companion was suffering terribly with his foot.'

"This conversation happened at Cox's railroad station [now center of Bel Alton], about a mile from Cox's house and young Mr. Cox said: 'Undoubtedly, Pa was in the underground railroad business, and so was Dr. Mudd. There was a Rebel post office in sight of this spot where you stand.' At this time the gentleman took me aside and pointed to a little bluff on the railroad cutting. 'Just at that spot,' said he, 'there was a stump in the midst of a dense woods which then covered this country, and that was the place where letters were deposited, either to be driven through to Richmond by runners or to be found by such people here as were familiar with the secret.' Some of the neighbors afterward told me that within a few yards of that stump was the hiding place of Booth and Harold [on side of today's Wills Road]. The general belief in that quarter is that Mr. Cox connived at their hiding in his woods, while James [Thomas A. Jones, no doubt], the blockade runner aforesaid, made ready to put them across the Potomac. It is also believed . . . that the Cox family had told the story that Booth's bay horse was shot in Zekiah Swamp about a mile from Booth's hiding place because he neighed and was short of food . . . some told me in Port Tobacco that the gray horse of Harold never was killed and was sold some time after the war. Others say both horses were shot in the swamp . . . and that the saddles are buried . . . also that the fugitives crossed the river near an old farm called 'Brentfield' near Pope's Creek and there is a well authenticated story that they mistook the Virginia for the Maryland shore [sic] where they were transported several miles up river into Nanjemoy Creek, where they went to the home of a Mr. [John J.] Hughes who knew Harold and begged something to eat. Two or three days later a frightened [about the Herold-Booth visit] Hughes went to Port Tobacco and gave information to his lawyer . . . "

(The lengthy remainder of the *Enquirer* article dealt with other aspects of the GATH journey.) [48]

George Alfred Townsend's 1883 account of the Booth-Herold sojourn in Maryland doubtless received coast-to-coast exposure. Because he was a syndicated columnist his treatment

---

[48] *The Cincinnati Enquirer*, Volume XLI, Number 177, June 26, 1883, p. 1, "GATH."

of the subject may well have been the most publicized, detailed account ever to appear in print. Apparently, Townsend in June 1883 did not know for certain about the important role played by Thomas A. Jones in the Booth-Herold escape. However, his journalist's nose and good sense did guide him directly to someone who knew a great deal about the Jones story in connection with his support of the Confederate cause.

A letter written September 18, 1883 by Townsend (from 361 West 34th St., New York City) to Mr. Thomas A. Jones in Baltimore follows:

> *"Dear Sir. Some weeks ago I visited Judge Stone near Port Tobacco and he described crossing the Potomac with you during the war and advised me to see you. I am a writer for the press and, sometimes, of books. It might be of mutual advantage for us to meet. If this reaches you and you will communicate with me I will speak further on the subject.*
>
> *Respectfully,*
>
> *Geo. Alfred Townsend"*

So began a series of notes from Townsend to Jones that continued for about six months.

And who was "Judge Stone," the man who blew the whistle on Thomas A. Jones? The Judge would not have mentioned the Jones name to a journalist in connection with the John Wilkes Booth escape, especially one with Townsend's connections, had he felt that doing so would harm Jones or his family in any way. He doubtless thought the time was ripe--and safe--for Thomas Austin Jones to come out of the closet and describe his role in helping Booth and Herold get across the Potomac and out of Maryland under the very noses and sabers of hundreds of patrolling Federal cavalry. Stone's comments about crossing the Potomac with Jones during the war indicated the judge knew a lot about the involvement of this Pope's Creek ferryman as a principal Confederate communications link during the conflict. It must have occurred to the Judge more than once as he served in the defense counsel at the June 1865 conspiracy trial that he might easily have found himself defending his old Confederate friend from Pope's Creek at the same time. It has been clear for years that Jones escaped the hangman's noose in July by not much more than a strand of the hangman's noose. In fact, it appears that Judge Frederick Stone himself may have played some significant role in supporting the Confederate cause as a prominent southern Maryland barrister with an impressive family and political background probably unmatched there at the time. His father, Frederick Daniel Stone, also a distinguished Port Tobacco lawyer in the 1820's and 1830's, was a son of 1st Judicial District Chief Judge the Honorable Michael Jenifer Stone. And he was a younger brother of Maryland Signer the Honorable Thomas Stone of "Habre de Venture" of Port Tobacco. Frederick Stone, orphaned in his teens, was guided through education and into the law by his distinguished uncle, William Briscoe Stone, also of Habre de Venture and another Port Tobacco lawyer. Surely it follows that

Judge Stone's giving away the Thomas A. Jones-Booth escape story could not have been considered and cannot be seen now as anything less than appropriate and honorable. So, the judge played the pivotal role in edging Jones out of the shade of anonymity when the timing seemed right and safe for all concerned.

There are five other original communications from Townsend to Jones written in the 1880's. The first, dated October 21, 1883, was sent to 69 Biddle Street, East of Gay Street in Baltimore, announcing, *"I will be at Barnum's Hotel early Friday evening."* Then follows an undated message, probably late 1883, to Mr. Thomas A. Jones, Corner of Wolfe and Eager Streets, Baltimore: *"Dear Sir, Can you reply by bearer whether you will call on me at Barnum's Hotel after dinner, say from six to eight o'clock. I am at the hotel now and have just arrived, G. A. Townsend."* Next, a lengthier note dated March 20, 1884, probably written in New York:

> *"... I send you today a magazine [Century- April issue] containing an article of which you are the hero ... I would not, I think, allow myself to be drawn into any long talks on this subject. The reporters in Maryland did not do themselves justice nor you the respect to tell your story and are not entitled to much consideration. I regret that I did not get a better look at you before I described you: for I see that you are a better looking man than I, at first, got sight of. However, make allowance for haste and imperfect understanding and believe me, your friend, Geo. Alfred Townsend."*

The *Century Magazine* article mentioned was an extremely interesting one done in great detail during the early spring of 1884. It incorporated most of the vital details of the Jones story pried from him by GATH during several 1883-84 Baltimore visits between the two men. [49]

In a communication dated April 8, 1884, the columnist wrote Jones,

> *"Sometime ago you thought you might be able to take a trip at my expense to your old resorts in Charles County. Could you go next Monday or Tuesday? I should like to go from La Plata to Pope's Creek and Allen's Fresh and cross over to Virginia. I received your letter and do not see how any considerate person could take offense at what was printed* [in 1884]. *I would be glad to hear from you on the subject of the journey. I have not asked Mr. Harbin to communicate with Mr. Stewart. Your friend, Geo. Alfred Townsend."*

---

[49] *The Port Tobacco Times and Charles County Advertiser* featured in its entirety the George Alfred Townsend 1884 article that disclosed the Thomas A. Jones' connection with Booth and Herold. *Times* issues of April 18, 25 and May 2, 1884 carried the lengthy story that first appeared in the April *Century Magazine*.

(Thomas H. Harbin was formerly a Confederate signal agent, postmaster at Bryantown, Charles County, and a relative of Jane Harbin Jones. Dr. Richard Stewart, one of the wealthiest men in King George County, Virginia, had totally supported the Confederate cause from beginning to end and was visited by Booth and Herold April 23, 1865 because Booth's ankle, treated by Samuel A. Mudd over a week earlier, was still extremely painful. Booth's visit to Dr. Stewart proved totally unrewarding.)

On April 12, 1884 Townsend sent a Western Union message to Jones saying that

> "... he would be in Baltimore Monday morning at Barnum's Hotel
> ... If you are ready, come there and we will take the train for
> Pope's Creek."

Whether or not this trip materialized is not known.

Townsend's first major effort to pin down details of the Booth-Herold escape trail in southern Maryland had begun in April 1883. The writer's terse penned diary notes that initiated his account of this journey begin on the page for April 10, 1883. [50]

> "Started 11:30 a.m. with rig for Surrattsville, Dr. Mudd's house
> and Bryantown, arrived latter place 5 p.m. Spent evening with Dr.
> G.D. [George Dyer] Mudd's house. Stopped at Bryantown Hotel
> [Murray's]."

Then, for April 11:

> "Started 8 a.m. to Bryantown Catholic Church, returned to
> Bryantown and proceeded to Beantown, Piscataway, to
> Surrattsville again and Washington, arrived at 2 p.m."

April 14: "This is the 18th anniversary of Lincoln's death - three days since over Booth's route."

June 1, 2, and 3: "Went today at 10 a.m. to see Frederick Stone [Judge] at Idaho, his seat two miles [east] of Port Tobacco, Md. Met him on train at Annapolis Junction and met his family at his house where I slept, talking with him about Mrs. Surratt, Herold and Dr. Mudd [Samuel A.] for whom he was counsel [at Lincoln conspiracy trial]. Went in a carriage with Mr. Stone to Coxes house and station [now Bel Alton] and to St. Thomas's Manor [St. Ignatius Church] and ate there with two Jesuit priests.

---

[50] These *GATH Diary Notes* are held by Maryland Archives in Annapolis, MD.

> *This morning, June 3, went to court house at Port Tobacco and met several young gentlemen of neighborhood and as I looked over the records [in the courthouse] and in p.m. drove to Habre de Venture, seat of Thomas Stone, the Signer, now owned by Miss Margaret Stone and to Rose Hill, seat of Dr. G.S. (sic) [Gustavus Richard] Brown, Washington's doctor now owned by Miss Olivia Floyd.*

[GATH's Charles County visit in early June was the one that led him to Thomas A. Jones, then living in Baltimore and not long since married again to Margaret Rountree and father of a three-year-old son]

> October 5: *"Talked to Thomas A. Jones in Baltimore who hid Booth and got him over the Potomac."*

> December 11: *"Today went to Baltimore and asked Thomas A. Jones to come to Barnum Hotel where I gave him $60 for his story of sending Booth to Virginia. We sat in [room] #52 until nearly midnight."*

> December 12: *"Today arrived in D. C. at 9 a.m. and looked over old maps of Virginia and returned to Baltimore and saw Jones at his coal yard, corner Eager and Wolfe Streets till 10 p.m. Took 11:35 train for New York."*

Townsend arrived back at his New York office with all the details he needed to shape his definitive account of what happened to Booth and Herold during the mysteriously unaccounted for final days of missing links in the April 1865 assassin escape to Virginia by way of southern Maryland. This story nearly a century and a half later leaves most people who know about it unable to comprehend fully the reason for such a grand sacrifice by one small farmer-fisherman. Tom Jones who doubtless would have passed through life without causing a ripple, had not a great man's murder violently upset whatever the fates might normally have had in store for him.

Century Magazine paid GATH $160 for the April 1884 Booth-Jones article; not a generous compensation when we consider what GATH went through to gather his information. But the writer indicated no dissatisfaction with his monetary settlement, according to a very brief comment in his diary entry for December 12, 1883.

Three of America's best known men of letters as they were on February 7, 1871. Left to right - George Alfred Townsend, Samuel L. Clemens and David Gray. Photo by Matthew Brady. Courtesy National Archives.

HOW WILKES BOOTH CROSSED THE POTOMAC
by George Alfred Townsend

The most dramatic of historical assassinations has had, until now, an unrelated interval. The actor John Wilkes Booth shot President Abraham Lincoln about ten o'clock Friday night April 14th, 1865. Near midnight he and his uninteresting road pilot, David E. Herold, called at Surratt's tavern, about ten miles south east of Washington, and obtained the arms, field glass, etc., previously prepared for them there. Saturday morning they were at Dr. Samuel A. Mudd's, twenty miles farther on, where Booth's broken ankle was set and a crutch made for him; and that evening the two fugitives were guided in a round-about way to the gate of Samuel Cox, a prosperous Southern sympathizer, about fifteen miles southwest.

The last witness in Maryland ended here. The Government, in its prosecution of the conspirators, took up the fugitive next at the crossing of the Rappahannock River in Virginia, on the 24th of April having failed to trace Booth a single step farther in Maryland, although he did not cross the Potomac until Saturday night, April 22nd. A whole week remained unaccounted for; and for the first time the missing links of the connection are here made public. Probably not half a dozen people are alive who have ever heard the narrative fully told.

When Annapolis was a greater place than Baltimore, and the Patuxent Valley the most populous part of Maryland, the main roads and ferries to all powerful Virginia were on the lower Potomac, instead of being, as now, above [near] Washington city. The most important of these ferries crossed at a narrow part of the river, where it is from two to three miles wide, near a stream on the Maryland side called Pope's Creek. Just below this spot, and not far above it, there are deep indentations from the river. A railroad, built since the war, for this reason has its terminus at Pope's Creek. About five miles north of the terminus is Cox's Station, which is about six miles south of the old courthouse village of Port Tobacco. A short distance east of Cox's Station is Samuel Cox's house; a short distance west of Cox's Station, perhaps two or three miles, is the old Catholic manor house of St. Thomas's, by an ancient church which gives the name to "Chapel" Point." Here the Potomac sends up the Port Tobacco River a broad tidal stream, naturally indicated at the beginning of the war as the nearest safe point for spies and go-betweens to reach broad water from Washington. Matthias Point on the Virginia side, makes a high salient angle into the waters of Maryland here, and is almost in the direct line from Washington to Richmond.

In this old region of the Calvert Catholics, a civilization existed at the close of the last century probably comparable with that of tobacco, and large landed estates, with slaves, were features of the high bluff country, which was plentifully watered with running streams amidst the hills of clay and gravel. But the Revolution emancipated the Catholic worship originally planted on the lower Potomac by the founders of Maryland, and a curious English society took root, with its little churches surmounted by the cross, its slaves attendant upon mass and confession; and much of the country, originally poor, was covered with decaying estates, old fields grown up in small pines, and deep gullies penetrating to the heart of the hills. The malaria almost depopulated the little towns and hamlets, tobacco became an uncertain crop, slavery kept the people poor, and intercourse fell off with the rest of the world, possibly excepting some of the old counties in Virginia in Washington's "Northern Neck."

Soon after the year 1820 Mr. [Samuel] Cox was born in the district below Port Tobacco, and his mother dying, he was put to nurse with Mrs. Jones, the wife of a plain man, possibly an overseer, who inhabited the house. She had a son, Thomas A. Jones, who grew up with young Cox; they were playmates and attended the same log school house, and Cox, as life progressed had the ruling influence over Jones, who was a cool, brave man, but without the self assertion of his comrade who soon developed into one of the most energetic men in that region.

A portrait of Samuel Cox shows him to have been of an indomitable will, strengthened by that consumptive tendency which often gives desperation to men fond of life. At the breaking out of the war Mr. Cox had thirty to forty slaves, plenty of land, a large house with out-buildings, Negro quarters, woodlands, and a superior appearance for these parts. He became captain of a volunteer company, which he drilled at Bryantown, a small settlement in the eastern part of the county, where the lands are unusually good and the neighbors plentiful in slaves. Hardly one of them an original secessionist, the course of events forced most of those slave holders into sympathy with the South, if not through their sensitiveness about their slave property, yet from the fact that their sons often hastened to cross the Potomac into the Confederate army, while in many cases their Negroes slipped off in the opposite direction within the Federal lines. The responsibility for disloyalty did not rest with these humble people off the great highways of life, but followed from the political consequences of breaking the Union asunder and leaving them on the Union frontier with all the necessities and traditions of slavery. The Government paid but little attention to them, seeing that they were below the line of military operations, divided by a broad river from the ragged peninsulas of the rebellion; and therefore, there almost immediately sprang up in lower Maryland, a system of contraband travel and traffic, which soon demoralized nearly everybody.

Thomas A. Jones, who had somewhat risen in the world and had a few slaves, sympathized warmly with the South; he owned a farm right at Pope's Creek, the most eligible situation of all for easy intercourse with Virginia. His house was on a bluff eighty to one hundred feet high, from which he could look up the Potomac to the west, across Mathias Point, and see at least seven miles of the river way, while his view down the Potomac was fully nine miles.

The moment actual war broke out, and intercourse ceased at Washington and above it with Virginia, great numbers of people came to Jones and to his next door neighbor on the bluff, Major Roderick G. Watson, asking to be sent across the Potomac. These fugitives were of all descriptions: lawyers, business men, women, resigned army officers, adventurers, suspected persons,--even the agents of foreign bankers and of foreign countries.

Major Watson had a large frame house, relatively new, two stories high, with dormer windows in the high roof, and with a servants' wing. He had a son in the Confederate army and grown up daughters; and his house became the signal station for the Confederates across the river, one of his daughters setting the signal, which consisted of a shawl or other black object, put up at the dormer window whenever it was not safe to send the boat across from Virginia. This window was kept in focus from Grimes's house on the other side, about two miles and a half distant--a small low house, planted at the water's edge from which the glass could read the signal, which no Federal officer, whether in his gun-boat or ashore, could suspect. Major Watson was somewhat advanced in years, and died while his neighbor Jones was serving an imprisonment in the Old Capitol Prison [in

Washington, D. C.]

On Jones's return to his home, he therefore became the most trusted neighbor of the Watson family, and they accommodated him as he assisted them. The young lady in the family was as enthusiastic for the Confederate cause, and as discreet in all her talks and walks as Jones himself, on whose countenance no human being could ever read what was passing within his mind. He had attended to his fishery and his farm until the war broke out, without having had an incident to mark his life: but suddenly there was an incursion of strangers to whose needs his rooted ideas of hospitality, no less than his sympathy for the Confederates, led him to hearken. His farming was almost broken up, and he took to crossing the river nearly every night, and sometimes twice or more of a night, with boats, sometimes rowed by two pairs of oars, at others by three, while he steered with an oar in the stern. The interlopers could ride down from Washington to Pope's Creek in six or seven hours, and Jones could put them at Grimes's house opposite in less than an hour. The idea of making money in this traffic never seems to have occurred to the man at all; he regarded these strangers as intrusted to his care by Providence or pity; and although his liberty was constantly in danger, he seldom received more than a dollar or two for taking anybody across. Some persons argued with him that he did not charge enough, and told him to look out for his family and the future; but, as the sequel will show, he did a vast amount of hard and dangerous labor for next to nothing, and in the end the Confederate Government also left him unpaid.

The original rebel route from Pope's Creek to Richmond was through Fredericksburg; but this being considerably to the west, a new route was opened over the old road to Port Royal on the Rappahannock River. Adventurers were taken by Jones or his neighbors across to Grimes's, who, assisted by one or two of his neighbors, carried them by vehicles in three or four hours to Port Conway, where a ferry was maintained across the Rappahannock River to Port Royal, and eighteen miles beyond it the high road from Washington to Richmond was open. Mr. Jones says that he may have crossed the Potomac one hundred times before he was arrested, but has no record of the days.

In the latter part of June, 1861 [sic October 1861], *General Sickles came with troops to the lower Potomac to keep a watch on the contraband intercourse. Grimes was found on the Maryland shore and sent to Fort Delaware. Jones was arrested when he returned from his visit to Richmond and sent to the Old Capitol prison at Washington and kept there six months. He was allowed to write to his family, subject to the inspection of his letters, and to talk to any of them when an officer was by. This imprisonment, together with his adventurous cruises previously, sharpened his wits, increased his knowledge of men and the world, and educated him for the official position he was soon afterward to occupy of Chief Signal Agent of the Confederacy North of the Potomac. Misfortunes, however, attended his affairs. His wife, who had a large family of children, was taken sick through care and confinement while he was absent, and died [in 1863]. His farm was mortgaged, and, not pursuing the regular vocations of peace, the mortgage slowly ate up the farm, and near the close of the war he had to remove from his river side residence to an old place called Huckleberry, about two miles and a half inland.*

Mr. Jones was released in March, 1862, by a general jail delivery ordered by Congress under the belief that the prisons were full of innocent men. He took an oath that he would not communicate with the enemy again, and was informed of the penalty of breaking it. He returned to his house on the river bluff, and soon an armed patrol and steam vessels were maintained on the river, and the Federal officers boasted that they had a spy on every farm. One of the fine old mansions on the river, Hooe's house, which had been the almost immemorial ferry house, was set on fire by the Federal flotilla and burnt for having given harborage to one of Grimes's boat parties.

Grimes again communicated with Jones, and asked him to go into an undertaking to carry the Confederate mail from Canada and United States to Richmond. Jones replied that the risk was too great, and that his duty to his children required him to stay home, although his heart was in the Confederate cause, and he would give it any assistance possible. Upon this, the Confederate signal officer, Major William Norris, who had been a Maryland man and is still alive, held an interview with Jones, and asked him to take charge of the rebel communications, stating that they were of the utmost consequence to the management of the Confederate cause and its intercourse with the outer world, the Federal blockade now being well maintained and every portion of the border closely watched, while the broad Potomac River and the pine covered hills of lower Maryland offered almost a sure crossing place. Finally Jones said that if he were given absolute control not only over the

Residence of Thomas A. Jones about a mile from the Potomac River near Pope's Creek from about the end of 1863 until about 1873, called then and now, "Huckleberry."

ferry, but over all agents to be retained in Maryland, the names of none of whom he should be called upon ever to mention he would undertake the work. He said to a Confederate agent: "It is useless to expect me to maintain a boat service with you. You must keep the boat on the Virginia side, cross to my beach, and bring and take the mail there, so that I cannot be suspected." He then indicated a post office in the hollow of an old tree which grew near the foot of his bluff.

His previous observations on the river had shown him that toward evening, when the sun had fallen below the Virginia woods, there was a certain grayness on the surface on the water, increased by the shadows of the high bluffs, which nearly erased the mark of a boat floating on the Potomac. The pickets that were now maintained along the bluffs were not set till toward night. Therefore it was arranged that the Virginia boat should come in just before the pickets were set, and its navigator noiselessly take out the mail from the old tree and deposit the Virginia packet, and then, with scarcely a word whispered or a sign given, slip back again to his Virginia cove. Generally, the boat was hauled ashore in Virginia out of the observation of the patrol gun boats and their launches, and sometimes it was kept west of Grimes's house, but sometimes back of Upper Machodoc Creek, which is six miles due south of Pope's Creek, only about twelve miles from Port Royal.

When the rebel mail had been left in the stump, Jones obtained it, either in person or by one of his faithful slaves. It is a singular fact that not only were women the best cooperative agents in this spy system, but the slaves, whose interests might be considered as opposed to a Southern triumph, frequently adhered to their masters from discipline or affection. Jones had a slave named Henry Woodland, still alive, who not only pulled in his boat to Virginia during the early months of the war, but, imitating the habits of his master, was discreet down to the time Booth escaped, while probably suspecting, if he did not know, all that was going on. He and his master seldom informed each other upon anything, and did not need even to exchange glances, so well did they know each other's ways. The Negro, nearly a duplicate of his master in methods, went about his work without speech, and asked no questions. Two other Negroes, named John Swan and George Murray, pulled oars in Jones's boats in the early part of the war. One of these, it is believed, turned spy and finally ran away, but was sent back to Jones by the commandant of the camp, received a flogging, and some time afterward deserted to a vessel in the river.

When the rebel mail had been put ashore, Jones would sometimes get it by slipping down through some of the wooded gullies cutting the bluff. The Federal patrol walked on the top of the bluff, and as the night grew dark would be apt to avoid these places, from which a shot might be fired, or an assassin spring. Jones sometimes ran risks getting down the bluff which was almost perpendicular, and after a time he constructed a sort of stairs or steps down one portion of it. His foster-brother Cox, who was more noisy and expressive, had contrived early in the war a set of post-offices for the deposit of the mail as it came up from the river, in stumps, etc. One of the post offices was pointed out to me where the railroad now goes through a cutting below Cox's Station. The Maryland neighbors, however, became so careless about sending their letters through these stump post-offices, that when Jones made his agreement with the Confederate Government, he dispensed with that system altogether, and depended upon more ordinary means. Feeling no passion for mere glory or praise, contented to do his work according to his own ideas of right and expediency, he merely made use of substantial, plain people, whose hearts were in the Confederate cause, but whose methods were all discreet. Thus he had a young woman to hoist his signal of black, and it never was hoisted if the course was open and clear on the river. He arranged that no mail matter should come close to his home, not even to Port Tobacco, that was perhaps ten miles distant. It was generally sent to Bryantown, fifteen to twenty miles distant, and collected there, or dispatched from that office, and it was carried by such neighbors as Dr. Stowton [Stoughton] W. Dent, who died in 1883, at the age of eighty. This old gentleman had two sons in the Confederate army, and was a practicing physician, riding on his horse from place to place, and it seemed to be the case that some persons in Major Watson's family was always sick. There the good old doctor would go, wearing a big overcoat with immense pockets, and big boots coming high toward his knees. Everybody liked him, the Federal officers and soldiers as well as the Negroes and neighbors, for he was impartial in his cares. At the greatest risk, even of his neck, the old man carried the rebel mail which Jones had delivered to him, and frequently went all the way to Bryantown with it. He would stuff his pockets, and sometimes his boots, with letters and newspapers.

There were one or two other persons sometimes made available as mail carriers. Perhaps Mr. Cox himself would do a little work of this kind. A man on the opposite side of the river, by the name of Thomas H. Harbin, who now lives in Washington, was a sort of general voluntary agent for the Confederacy, making his headquarters now in Washington and now in Richmond, and again on the river bank. In his desire to accommodate everybody, Harbin sometimes put too much matter in the mail and Jones's cautious soul was much disturbed to find, on one occasion, two large satchels filled with stuff not pertinent to the Confederate government. He sent word over that there must be more sense in the putting in of that mail, as it would be impossible to get it off if it grew larger.

Jones's house at this time was of dark, rain-washed plank, one story high, with a door in the middle, an outside chimney at each end, and a small kitchen and intervening colonnade which he added himself. The house was about thirty yards from the edge of the bluff. His farm contained five hundred and forty acres. Besides the Watsons below, Mr. Thomas David Stone had a place just above him, across Pope's Creek, on a high hill called "Ellenborough." Next above Stone's, was George Dent, who had an interesting mansion. The third farm, to the north was Brentfield, and back of it "Huckleberry," from which Booth departed [night of April 21].

Mr. Jones himself is a man of hardly medium height, slim and wiry, with one of those thin, mournful faces common to Tidewater Maryland with high cheek bones, gray-blue eyes, no great height or breadth of forehead, and thick, strong hair. The tone of his mind and intercourse is slow and mournful, somewhat complaining, as if the summer heats had given a nervous tone to his views, which are generally instinctive and kind. Judge Frederick Stone told me that he once crossed the river with Jones, when a Federal vessel suddenly loomed up, apparently right above them, and in the twinkling of an eye, the passenger said, he could see the interior of the Old Capitol prison for himself and all his companions; but at that moment Jones was as cool as if he had not noticed the vessel at all, and extricated them in an instant from the danger. Jones's education is small. He does not swear, does not smoke, and does not drink. When he was exposed on the river, he says, he sometimes took a little spirits to drive away the cold and wet; but he has few needs, and probably has not changed any of his habits since early life. Born poor, somewhat of the overseer class, and struggling toward independence without greed enough ever to accomplish it, he was eminently made to obey instructions and to keep faith. His neighbor Cox was more subtle and influential and although he was rough and domineering, seldom failed to bring any man to his views by magnetism or persuasion. Jones's judgment often differed from Cox's, and in the end his courage was altogether superior; but still, from early habits, the humble farmer and fisherman always yielded at last to what Cox insisted upon.

Mr. Jones was not alone in his operations during the war, but he was the only trusted man in Maryland with whom the Confederate Government had an official relation. His very humility was his protection. He impressed the Federal officers and Union men generally as a man of rather slow wits, of an indolent mind, with but little intelligence or interest in what was going on around him. Yet a cunning which had no expression but acts, a devotion which never asked to be appreciated, and perseverance to this day remarkable, were his. Some of his neighbors were running boats across the river for hire or gain. In the little village of Port Tobacco most of the mechanics and loungers had become demoralized by this traffic, and among these was George A. Atzerodt, a coach maker, of but little moral or physical stamina; who was afterward hanged among the conspirators. This man left his work after the war began, and took to the business of pulling a boat down Port Tobacco River to Virginia. Among the persons who occasionally crossed the river was John H. Surratt, a country boy of respectable aspirations until some time after the breaking out of the war, when he, too, was caught in the meshes of the contraband trade, and, possessing but little mind and too much vanity, was carried away with his importance. Jones went to Richmond once or twice toward the close of the war, and on one of these occasions Surratt and a woman under his care crossed in the same boat. Sometimes these boats would go so heavily laden that a gale on the river would almost capsize them. One portion of Jones's business was to put the New York and Northern newspapers every day into Richmond. These newspapers would go to Bryantown post office, or sometimes to Charlotte Hall post office, and generally reach the Potomac near dusk, and being conveyed all night by the Confederate mail carriers, by way of Port Royal, would be in the hands of the rebel Cabinet next morning, twenty-four hours only after the people in New York were reading them; and Jones says that there was hardly a failure one day in the year to take them through.

The Federal authorities never had a tithe of the thoroughness of suspicion and violation of personal liberty which the Confederates always exercised. Hence the doom of Abraham

Lincoln was slowing coming onward through these little countryside beginnings, starting without origin and ending in appalling calamity.

About the third year of the war, Jones understood that a very important act had been agreed upon, namely, to seize the President of the United States in the city of Washington, and by relays and forced horses take him to the west side of Port Tobacco Creek, about four miles below the town of that name, and dispatch him across the Potomac a prisoner of war. I possess the names of the two persons on Port Tobacco Creek who, with their sons, were prominent in this scheme; but the frankness with which the information was given to me persuades me not to print them. A person already named, in Washington, was in the conspiracy; and it was given out that "the big actor, Booth," was also "in it." Jones heard of this about December 1864. It was not designed that he should take any part in the scheme, though he regarded it as a proper undertaking in time of war, the bateau which was to carry Mr. Lincoln off was kept ready, and the oars and men were ever near at hand, to dispatch the illustrious captive.

That winter was unusually mild and therefore the roads were particularly bad in this region of clay and marsh, and did not harden with the frost--a circumstance which perhaps spared Mr. Lincoln the terrors of such a desperate expedition. Inquiries were made from time to time as to when the thing was to be done, and it was generally answered that the roads were too heavy to give the opportunity. The idea Jones has of this matter is that Mr. Lincoln was to be seized, not on his way to the Soldiers' Home, but near the Navy Yard, and gagged quietly, and the carriage then driven across the Navy Yard bridge or the next bridge above, while the captors were to point to the President and wave their hands to the guards on the bridge, saying, "The President of the United States." When we consider that he was finally killed in the presence of a vast audience, and that his captors then crossed the same bridge without opposition and without passes, the original scheme does not seem extraordinary. There is no doubt but that in this original scheme the late Dr. Samuel A. Mudd was to play some part. Booth had made his acquaintance during that fall or winter on his first visit to the county, and some of Dr. Mudd's relatives admit that he knew Booth well, and probably was in the abduction scheme. The calculation of the conspirators was that the pursuers would have no opportunity to change horses on the way, while the captors would have fresh horses every few miles and drive them to the top of their speed, and all they required was to get to the Potomac River, seven hours distant, a very little in advance. The distance was from thirty-six to thirty-eight miles, and the river could be passed in half an hour or little more with the boat all ready. Jones thinks that this scheme never was given up, until suddenly information came that Booth had killed the President instead of capturing him, and was supposed to be in that region of country. Jones had never seen Booth, and had scarcely any knowledge of him.

When Jones went to Richmond, just before the assassination, it was to collect his stipend, which he had confidently allowed to accumulate until it amounted to almost twenty-three hundred dollars, presumably for three years' work. He reached Richmond Friday, and called on Charles Caywood, the same who kept the signal camp in the swampy woods back of Grimes's house. The chief signal officer said he would pay five hundred dollars on Saturday, but if Jones would wait till Tuesday the whole amount would be paid him. Jones waited. Sunday night Petersburg fell, and on Monday Richmond was evacuated, so the Confederacy expired without paying him a cent. Moreover, he had invested three thousand dollars in Confederate bonds earlier in the war, paying for them sixty-five cents on the dollar, and keeping them till they were mere brown paper in his hands.

Jones heard of the murder of Lincoln on Saturday afternoon, April 15th, at or near his own farm of Huckleberry. Two Federal officers or cavalrymen came by on horseback, and one of them said to Jones, "Is that your boat a piece above here?" "Yes," said Jones. "Then you had better take good care of it, because there are dangerous people around here who might take it to cross the river." "That is just what I am thinking about," said Jones, "and I have had it pulled up to let my black man go fishing for the shad which is now running." The two horsemen conferred together a minute or two, and one of them said:

"Have you heard the news from Washington?" "No." "Our President has been murdered." "Indeed!" said Jones, with a melancholy face, as if he had no friend left in the world. "Yes," said the horseman, "President Lincoln was killed last night, and we are looking out for the men, who, we think, escaped this way."

On Sunday morning, the 16th of April, about nine o'clock, a young white man came from Samuel Cox's to Jones's second

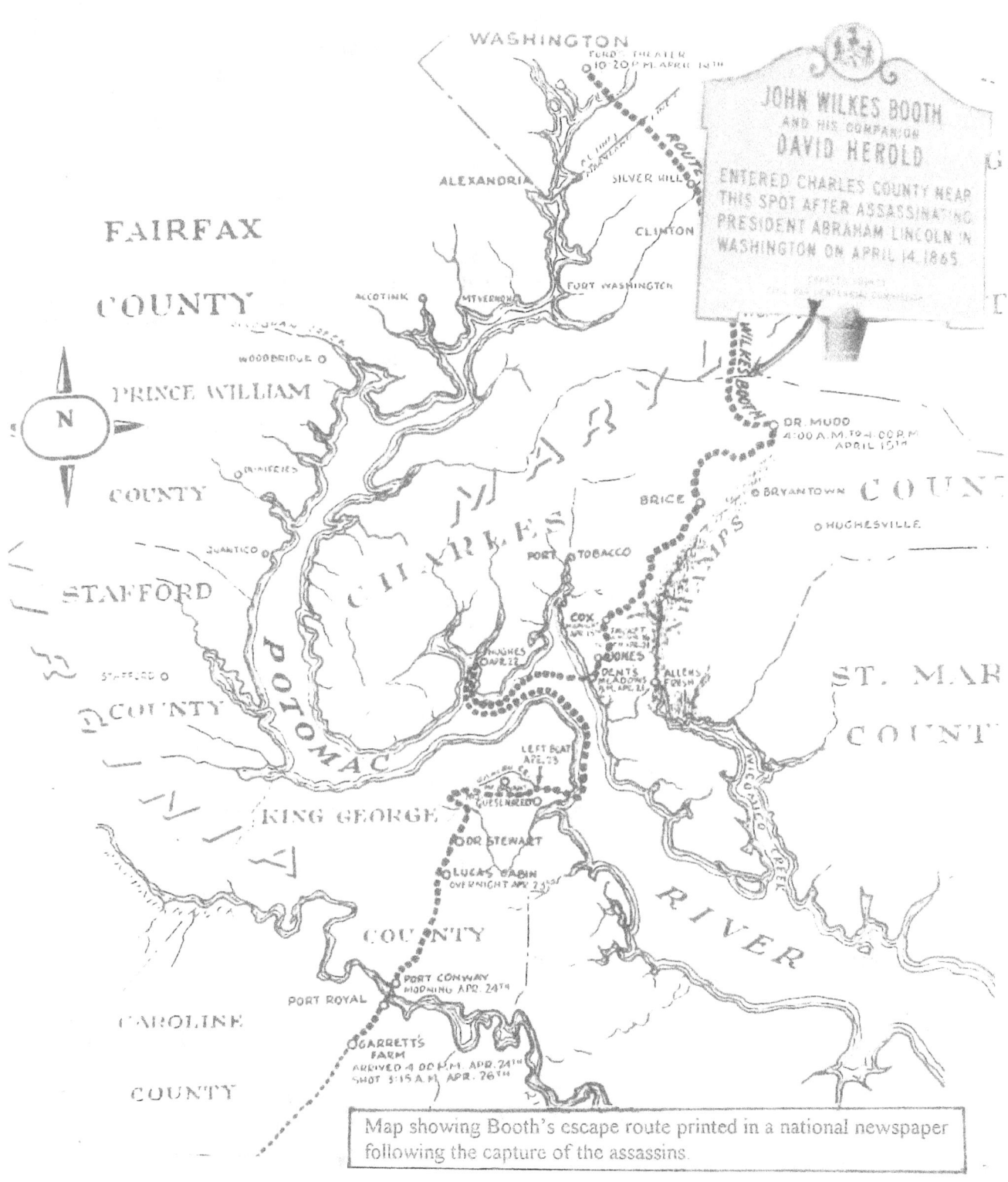

Map showing Booth's escape route printed in a national newspaper following the capture of the assassins.

farm, called "Huckleberry," which has been already described as about two and a half miles back from the old river residence, which Jones had been forced to give up when it appeared probable that the Confederate cause was lost. The "Huckleberry" farm consisted of about five hundred acres, and had on it a one-story and garret house, with a low pitched roof, end chimneys, and door in the middle. There was a stable north of the house, and a barn south of it, and it was only three quarters of a mile from the house to the river which here runs to the north to make the indentation called Port Tobacco Creek or river. Although Jones, therefore, had moved some distance from his former house, he was yet very near tide water. The new farm was much retired, was not on the public road, and consisted of clearings amidst rain washed hills with deep gullies, almost impenetrable short pines and some swamp and forest timber. Henry Woodland, the black servant, who was then about twenty seven years old, was still Jones's chief assistant, and was kept alternately farming and fishing.

The young man who came from Cox's was told, if stopped on the road, to say that he was going to Jones's to ask if he could let Cox have some seed corn, which in that climate is planted early in April. He told Jones that Colonel Cox wished him to come immediately to his house, about three miles to the north. The young man mysteriously intimated that there were very remarkable visitors at Cox's the night before. Accustomed to obey the summons of his old friend, Jones mounted his horse and went to Cox's. The prosperous foster brother lived in a large two story house, with handsome piazzas front and rear, and a tall, windowless roof with double chimneys at both ends; and to the right of the house, which faced west, was a long one story extension, used by Cox for his bedroom. The house is on a slight elevation, and has both an outer and inner yard, to both of which are gates. With its trellis work and vines, fruit and shade trees, green shutters and dark red roofs, Cox's property called "Rich Hill," made an agreeable contrast to the somber short pines which, at no great distance seemed to cover the plain almost as thickly as wheat straws in the grain field.

Taking Jones aside, Cox related that on the previous night the assassin of President Lincoln had come to his house in company with another person, guided by a Negro, and had asked for assistance to cross the Potomac River; "and," said Cox to Jones, "you will have to get him across." Cox indicated where the fugitives were concealed, perhaps one mile distant, a few rods west of the present railroad track, and just south of Cox's station. [The tracks were laid down in 1871.] *Jones was to give a signal by whistling in a certain way as he approached the place, else he might be fired upon and killed. Nobody, it is believed, ever saw Booth and Herold after this time in Maryland, besides Cox's overseer, Franklin Roby, and Jones. Cox's family protest that the fugitives never entered the house at all; his adopted son, still living, says Booth did not come in the house. Herold, who was with Booth, related to his counsel, as the latter thinks, that after they left Mudd's house, they never were in any house whatever in Maryland. The Negro* [Oswald Swann, who actually was a Native American] *who was employed to guide Booth from Dr. Mudd's to Cox's testified that he saw them enter the house; but as the Government did not use him in the trial, it is probable that he related his belief rather than what he saw.*

But there is no doubt of the fact that when Dr. Mudd found Booth on his hands on Saturday, with a broken ankle, and the soldiers already pouring into Bryantown, he and Booth and Herold became equally frightened and in the early evening the two latter started by a road to the east for Cox's house, turning [around] *Bryantown and leaving it to the north, and arriving about midnight at Cox's. There the Negro was sent back. Herold advanced to the porch and communicated with Cox, and Booth sat on his horse off toward the outer gate. The two men cursed Cox after they backed out to where the Negro was--he remaining at the outer gate--and said that Cox was no gentleman and no host. These words were probably intended to mislead the Negro when they sent him back to Dr. Mudd's. This Negro was arrested, as was a colored woman* [Mary Swann] *in Cox's family, and, with the same remarkable fidelity I have mentioned, the woman confronted the Negro man and swore that what he said was untrue.*

Nevertheless, Booth and Herold were sent into the short [scrub] *pines, and there Jones found them. He says that as he was advancing into the pines he came upon a bay mare, with black legs, mane, and tail, and a white star on the forehead; she was saddled, and roving around in a little cleared place as if trying to nibble something. Jones took the mare and tied her to a tree or stump. He then advanced and gave what he called the countersign or whistle, which he does not precisely remember now, though he thinks it was two whistles in a peculiar way, and a whistle after an interval. The first person he saw was Herold, fully armed, and with a carbine in his hand, coming out to see who it was. Jones explained that he had come to see them, and he was then taken to Booth, who was but a few rods farther along.*

Booth was lying on the ground, wrapped up in blankets, with his foot supported and bandaged, and a crutch beside him. His rumpled dress looked respectable for that country, and Jones says it was of black cloth. His face was pale at all times, and never ceased to be so during the several days that Jones saw him. He was in great pain from his broken ankle, which had suffered a fracture of one of the two bones in the leg, down close to the foot. It would not have given him any very great pain but for the exertion of his escape, which irritated it by scraping the ends of the bone broken, perhaps in the flesh; it was now highly irritated, and whichever way the man moved he expressed by a twitch or a groan the pain he felt. Jones says that this pain was more or less aggravated by the peril of Booth's situation--unable to cross the river without assistance, and unable to walk any distance whatever. Jones believes that Booth did not rise from the ground at any time until he was finally put on Jones's horse to be taken to the waterside some days afterward.

Booth's first solicitude seemed to be to learn what mankind thought of the crime. That question he put almost immediately to Jones, and continued to ask what different classes of people thought about it. Jones told him that it was gratifying news to most of the men of Southern sympathies. He frankly says that he himself at first regarded it as good news; but somewhat later, when he saw the injurious consequences of the crime to the South, he changed his mind. Booth desired newspapers if they could be had, which would convey to him an idea of public feeling. Jones soon obtained newspapers for him, and continued to send them in; and Booth lay, there where the pines were so thick that one could not see more than thirty or forty feet into them, reading what the world had to say about his case. He seemed never to tire of information on this one subject, and the only thing besides he was solicitous about was to get across the river into Virginia.

Jones says Booth admitted that he was the man who killed Lincoln, and expressed no regret for the act, knowing all the consequences it involved. He harped again and again upon the necessity of his crossing the river. He said if he could only get to Virginia he could have medical attendance. Jones told him frankly that he would receive no medical attendance in Maryland. Said he: "The country is full of soldiers, and all that I can do for you, if I can, for Cox's protection and my own, and for your own safety. That I will do for you, if there is any way in the world to do it.

When I [Gath] *received this account from Mr. Jones, I asked him question after question to see if I could extract any information as to what Booth inquired about while in that wilderness. I asked if he spoke of his mother, of where he was going when he reached Virginia, of whether he meant to act on the stage again; whether he blamed himself for jumping from the theater box; whether he expressed any apprehensions for Mrs. Surratt or his friends in Washington. To these and to many other questions Jones uniformly replied; "No, he did not speak about any of those things. He wanted food, and to cross the river, and to know what was said about the deed." Booth he thinks wore a slouched hat. At first meeting Booth in the pines, he proved himself to be the assassin by showing upon his wrist, in India ink, the initials J. W. B. He showed the same to Captain Jett in Virginia. Jones says Booth was a determined man, not boasting, but one who would have sold his life dear. He said he would not be taken alive.*

Mr. Jones went up to Port Tobacco in a day or two to hear about the murder, and heard a detective there from Alexandria say; "I will give one hundred thousand dollars to the man who can tell where Booth is." When we consider that the end of the war had come and all the Confederate hopes were blasted and every man's slaves set free, we may reflect upon the fidelity of this poor man, whose land was not his own, and with inevitable poverty before him perhaps for the rest of his days, when the next morning he was told [that] *to him alone would be intrusted that man for whom the Government had offered a fortune, and was increasing the reward. Mr. Jones says it never occurred to him for one moment that it would be a good thing to have that money. On the contrary, his sympathies were enlisted for the pale faced young man, so ardent to get to Virginia and have the comforts of a doctor.*

Said he to Booth; "You must stay right here, however long, and wait till I can see some way to get you out; and I do not believe I can get you away from here until this hue and cry is somewhat over. Meantime, I will see that you are fed." He then continued to visit them daily, generally about 10 o'clock in the morning. He always went alone, taking with him such food as the country had--ham, whisky, fish, bread, and coffee. Part of the way Jones had to go by the public road, but he generally worked into the pines as quickly as possible. His intercourse at each visit with the fugitives was short, because he was in great personal danger himself, he was not inquisitive, and was wholly intent on keeping his faith with his old friend and the new ones. He says that Herold had nothing to say of the least importance, and was nothing but a pilot for Booth. Not improbably, Cox sent his own overseer into the pines sometimes to see these men and to give them something, but he took no active part in their escape. The blankets they possessed came either from Cox's or from Dr. Mudd's.

Booth, as has been said, rode a very small bay mare from the rear of Ford's Theatre to Cox's pines. Herold rode a horse of another color. These horses were hired at different livery stables in Washington. Jones is not conversant with all the facts about the shooting of these horses, but the testimony of Cox before he died was nearly as follows: After Booth entered the pines he distinctly heard the next day or the day following, a band of cavalry going along the road at no great distance, and the neighing of their horses. He said to Herold; "If we can hear those horses, they can certainly hear the neighing of ours, which are uneasy from want of food and stabling." When Jones on Sunday morning came through the woods and found one of the horses loose, he told Cox, as well as Booth, that the horses ought to be put out of the way. Cox had Herold advised to take the horses down into Zekiah Swamp [about half a mile eastward] and shoot them both with his revolver, which he did.

The weather during those days and nights was of a foggy, misty character--not cold, but uncomfortable, although there was no rain. At regular intervals the farmer got on his horse and went through the pines the two or three miles to the spot where still lay the yearning man with the great crime behind him and the great wish to see Virginia. Booth had a sympathetic nature, and seldom failed to make a good impression; and that he made this impression on Jones will presently appear. No incident broke the monotony of these visits for days. Jones sent his faithful Negro out with the boat to fish will [sic] gill-nets, so that it should not be broken up in the precautions used by the Federals to prevent Booth's escape. Jones was now reduced to one poor boat, which had cost him eighteen dollars in Baltimore. He had lost several boats in the war, costing him from eighty to one hundred and twenty five dollars apiece. This little gray or lead-colored skiff was the only means by which the fugitives could get across the river. Every evening the man [Woodland] returned it to the mouth of the little gut or marsh called Dent's Meadow, in front of the Huckleberry farm. This is not two miles north of Pope's Creek, and from that spot Booth and Herold finally escaped. [Dent's meadow lay between Huckleberry and the river shore - now a marsh.]

Thomas A. Jones about 1890.

*Monday, Tuesday, Wednesday and Thursday passed by, and more soldiers came in and began to ride hither and thither, and to examine the marshes; but they did not penetrate the pines at all, which at no time were visited. The houses were all examined, and old St. Thomas's brick buildings, of a venerable and imposing appearance, above Chapel Point, were ransacked. The story went abroad that there were vaults under the priests' house, leading down to the river, and finally the soldiers tore the farm and terraces all to pieces. Yet, **for six nights and days** Booth and Herold kept in the woods. On Friday Jones slipped over to a little settlement called Allen's Fresh, two or three miles from his farm, to see if he could hear anything. A large body of cavalry were in the little town, guided by a Marylander, and while Jones in his indifferent way was loitering about he heard the officer say; "We have just got news that those fellows have been seen down in St. Mary's county." The cavalry were ordered to mount and set out. At that time it was along toward the gray of the night, and instantly Jones mounted his horse and rode from Allen's Fresh by the road and through the woods to where Booth and Herold were.*

Said he with decision; "Now, friends, this is your only chance. The night is pitch dark and my boat is close by. I will get you some supper at my house, and send you off if I can." With considerable difficulty, and with sighs and pain, Booth was lifted on to Jones's horse, and Herold was put at the bridle. "Now," whispered Jones, "as we cannot see twenty yards before us, I will go ahead. We must not speak. When I get to a point where everything is clear from me to you, I will whistle so," giving the whistle. In that way he went forward through the darkness, repeating the signal now and then; and although the wooded paths are generally tortuous and obstructed, nothing happened. For a short distance they were on the public road; they finally turned into the Huckleberry farm, and about fifty yards from the house the assassin and his pilot stopped under two pear trees.

At this moment a very pathetic incident took place. Jones whispered to Booth; "Now I will go in and get something for you to eat, and you eat it here while I get something for myself." Booth, with a sudden longing, exclaimed; "Oh, can't I go in the house just a moment and get a little of your warm coffee?" Jones says he felt the tears come to his eyes when he replied; "Oh, my friend, it would not be safe. This is your last chance to get away. I have Negroes at the house; and if they see you, you are lost and so am I." But Jones says, as he went in, he felt his throat choked. To this day he remembers that wistful request of the assassin to be allowed to enter a warm habitation once more before embarking on the wide and unknown river.

Henry Woodland was in the kitchen taking his meal, and neither looking or asking any question, though he must have suspected from the occurrences of a few days past that something was in the wind. "Henry," said Jones, "did you bring the boat back to Dent's meadow where I told you?" "Yes, master." "How many shad did you catch, Henry?" "I caught about seventy, master." "And you brought them all here to the house, Henry?" "Yes, master."

Jones then took his supper without haste, and rejoined the two men. It was about three quarters of a mile to the water side, and, although it was very dark, they kept on picking their way down through the ravine, where a little, almost dry stream ran off to the marshes. Not far from the water side was a strong fence, which they were unable to take down.

Booth was now lifted from the horse by Herold and Jones, and they got under his arms, he with the crutch at hand, and so they nearly carried him to the water. The boat could be got by a little wading, and Jones brought it in. Booth took his place in the stern. He was heavily armed, and Jones says had not only his carbine, as had Herold, but revolvers and a knife. Herold took the oars, which had been left in the boat, and sat amidships. Jones then lighted a piece of candle which he had brought with him, and took a compass which Booth had brought out from Washington, and by the aid of the candle he showed Booth the true direction to steer. Said he; "Keep the course I lay down for you, and it will bring you right in Machodoc Creek. Row up the creek to the first house, where you will find Mrs. Quesenberry, and I think she will take care of you if you use my name."

They were together at the water side an unknown time, from fifteen minutes to half an hour. At last Booth with his voice full of emotion, said to Jones; "God bless you, my dear friend for all you have done for me." The last words Jones thinks Booth said were; "Good bye, old fellow!" There was a moment's sound of oars on the water, and the fugitives were gone.

For the danger and the labor of those six days Jones received from Booth seventeen dollars in greenbacks, or a little less than the cost of the boat which Jones had to surrender forever. Booth had about three hundred dollars in his possession, and he told Jones that he was poor, and intimated that he would give him a check or draft on some one, or on some bank. "No," said Jones; "I don't want your money. I want to get you away for your own safety and for ours."

It was not until months after this that Jones ascertained that the fugitives did not succeed in crossing the river that Friday night. They struck the flood tide in a few minutes, were inexperienced in navigating and when they touched the shore sometime that night and discovered a house near by, to which Herold made his way, the latter saw something familiar about the place, he knowing all that country well. It was the residence of Colonel John J. Hughes, near Nanjemoy Stores, in Maryland, directly west of Pope's Creek, about eight or nine miles. [Nanjemoy Stores actually was a country general store on the Potomac River at Riverside. It was three miles from the east side of the Nanjemoy Creek landing place near the Hughes house.] *The Potomac here is wide, and has many broad inlets. In the darkness the Virginia shore and Maryland shore seemed the same. Herold went to the house and asked for food, and said that Booth was in the marsh near by, where they had pulled up the boat out of observation. The good man of the house was much disturbed, but gave Herold food, and is supposed that after being concealed that day they pushed off again in the evening, and this time successfully made the passage of the river, though they had to come back twelve to fourteen miles. The keeper of the house at Nanjemoy became frightened after they left, and rode into Port Tobacco and told his lawyer of the circumstance who took him at once before a Federal officer.*

Some time on Sunday morning, the ninth morning after the assassination, the fugitives got to Machodoc Creek, at Mrs. Quesenberry's with whom they left the boat. It is not sure that they entered her house, but they went to the house of a man named Bryan on the next farm, and probably revealed themselves. Bryan next day took them to the summer house of Dr. Richard Stewart, which is two or three miles back in the country. This Dr. Stewart was the richest man in King George county, Virginia, and had a very large brick house at Mathias Point on the river; but on account of the malaria and heat he went in summer to a large barn-like mansion back in the woodlands, a queer, strange house two stories high, with a broad passage. He was entertaining some friends just returned from the Confederate service, and much annoyed to find that on his place were the assassins of President Lincoln, after the war was all over. The men were not invited into the house, but were sent to an out building of some kind, either the Negro quarters or the barn; and Booth was so much chagrined at this welcome to Virginia that he took the diary which was found on his dead body and wrote a letter in lead pencil to Dr. Stewart, sorrowful rather than angry, saying that he would not take hospitality extended in that way without paying for it, and sending three dollars.

Booth procured a conveyance, or one was procured for him, from Dr. Stewart's to Port Conway; it was driven by a Negro named Lucas. He probably spent Sunday in Bryan's house, and got to Dr. Stewart's house, it is said, on Monday, where he asked for breakfast, and the same day reached the Rappahannock River and went across with Captain Jett. This crossing was made on Monday, the twenty-fourth of April. That afternoon he was lodged at Garrett's farm three miles back. He spent the next day at this house and slept in the barn. Being informed that a large body of Federal cavalry had gone up the road this Tuesday, he became much distressed. On Wednesday morning, soon after midnight, the cavalry returned, guided by Captain Jett. The barn was set afire and Booth shot soon after 3 o'clock in the morning. He died a little after sunrise on Wednesday [while stretched out on the porch of the Garrett house].

I may recapitulate Booth's diary during those days as Jones has indicated it. At ten o'clock Friday night Booth shot the President. A little after midnight he was at Surratt's tavern, where he received his carbine and whisky. (I forgot to say that, among the articles of comfort given to Booth by Jones when he went to the boat, was a bottle of whisky.) In gray dawn of Saturday morning Booth was at Dr. Mudd's, where he had his leg set, and a laboring white man there whittled him a crutch. On Saturday night, near midnight, he was at Cox's house, and some time between that and morning was lodged in the pines, where he remained Sunday, Monday, Tuesday, Wednesday, Thursday and Friday; and Friday night, between eight and nine o'clock, he started on the boat, spent Saturday in Nanjemoy Creek, and arrived sometime Saturday night or before light on Sunday at Mrs. Quesenberry's It is understood that on the Virginia side he was welcomed by two men named Harbin and Joseph Baden, the latter of whom is dead. The boat in which Booth crossed the river, he gave Mrs. Quesenberry, who was arrested. The boat was put on a war vessel and probably carried to Washington.

A few days after Booth crossed the river and had been killed, suspicion turned upon both Jones and Cox. The Negro who had taken the fugitive to Cox's gate gave information. Negroes near Jones's farm said he had recently concealed men, and showed the officers a sort of litter or camp about two hundred yards from his house. Here, in reality, quite a different fugitive had hidden sometime before. Jones looked at it in his mournful way, and expressed the opinion that it was nothing but where a hog had been penned up. He was arrested and taken to Bryantown, and kept there eight days in the second story of the tavern where Booth had stopped, and in sight of the country Catholic church [St. Mary's] where Booth first met Dr. Mudd and others, six months before. Cox was there, but was in two or three days sent to Washington. The detectives from all the cities of the East sat in the street under Jones, and described how he was to be hanged. He remarks of Colonel Wells; "He was a most bloodthirsty man and tried to scare out of me just what I'm tellin' of you now." In eight days Jones was sent to the old Carroll prison, Washington. There he contrived to communicate with Cox, who was completely broken in spirit, and told him by no means to admit anything; and when Jones, in about a month, saw Swan, the Negro [Indian] witness, going past his window toward the Navy Yard bridge with a satchel, Jones said to Cox; "You have nothing to fear." The Government soon released these then, who indeed had taken no part in Mr. Lincoln's death, though they may have been accomplices after the fact. Jones was kept six and Cox seven weeks.

He has filled some places under the Maryland and Baltimore political governments, and now keeps a coal, wood and feed yard in North Baltimore. [At corner of Wolf and Eager Streets.]"

Townsend's remarkably insightful and gracious obituary in 1899 about Charles County's distinguished, post Civil War jurist, Frederick Stone, who had died on the 7th of October at his La Plata home (Idaho), gives us today an intimate look at the make-up of southern Maryland society as experienced by Townsend himself. GATH wrote that,

". . . his ways were what he called 'old style' . . . but his head and feet were in the living, practical present . . . a few such wise, cautious and learned men could have revived the district in which

resistance; he said it was a profound mistake . . . " [51]

Judge Stone's Charles County social and professional contemporaries who knew him very well backed up the GATH obituary appraisal of the judge as man and jurist. A Maryland Court of Appeals "Memoriam" to Stone on December 9, 1899 includes singular praise for Honorable Frederick Stone who had been a highly respected member of this high court for years until forced to retire because of a statutory age limit on service. Charles County's distinguished and politically well-connected attorney L. Allison Wilmer delivered an impressive eulogy before the Court. After reading the Court-composed Memorial he added the following words extemporaneously:

> *". . . the charm and grace of his character lay in his freedom from malice and ill-will . . .* [he] *had a magnanimity, a spirit of forgiveness, a charity most rare among men. He grew old beautifully . . . the rougher edges which we frequently find in men of honest, blunt, robust character, had worn away, and the native gentleness and kindliness of his disposition shone as the polished corners of the temple."*

Court of Appeals Chief Judge McSherry said next that Stone's vigorous mind...

> *"allowed him to grasp intuitively the salient points of a controversy and dispose of them concisely, in simple and terse language."*

This good man had for years been mentor, friend and protector of Thomas A. Jones. Perhaps each knew more about the other's Civil War experiences than anyone else could ever know. [52]

And the Townsend-Stone relationship was the primary instrument for bringing Thomas A. Jones to public view in a way that did him no harm and, indeed, brought him a broad measure of public respect and sympathy. But, as always there was no significant hard-cash compensation for Tom Jones. He was forced to return to Charles County in 1887, desperately poor.

---

[51] *Op cit*, GATH Stone Obituary.

[52] John M. Wearmouth, *Charles County Helps Shape the Nation*, (La Plata, Maryland: 1988), pp. 33-35.

PART 5

INTRODUCTION TO BOOK BY THOMAS A JONES

J. WILKES BOOTH

Copyright, 1893, by J. B. Mattingly
Chicago
Laird & Lee, Publishers

    This rather slim work by Jones has been in and out of print at least three times since 1893. It is being reproduced here to help it remain in print and to make sure that the life of Thomas Austin Jones is not in any way overshadowed by the broader story of the total John Wilkes Booth-David Herold escape through Maryland. Earlier works by Townsend printed in periodicals synthesized much relevant material, including details that even Jones himself did not know. Town send's reporting was professionally done and resulted from an incredibly impressive amount of research. However, the Thomas A. Jones's presentation includes some different slants based on his intimate connection with the Lincoln assassination fugitives in Charles County. And, doubtless, the leveling and mellowing influence of the years following the Town send-Jones disclosures in the 1880's also colored the 1893 narrative. Too, its telling probably was shaped in part by the avid and expert ghost-writing of La Plata lawyer John E. Stone, who had been a neighbor and friend of Jones throughout Stone's life. His father, Thomas David Stone of "Ellenborough" at Pope's Creek, doubtless knew as much as anyone during the Civil War about Jones's handling of Confederate secret service matters and mail to, through and within the greater Pope's Creek communication channel-funnel. Because this narrative was produced primarily to be a money-raising feature at the Columbian Exposition in Chicago it should have taken into account the location and temper of the times at that place. Jones knew that he was going into Abraham Lincoln's back yard and that thousands of Midwestern Civil War veterans would be visiting the great exposition and many of them could be offended by apparent profiteering related directly to assistance given the assassin of Lincoln. Resentment materialized to the point where Jones, and J. Benjamin Mattingly who had accompanied him, had to pack up and return home . . . probably much the poorer for their effort.

    The plan to sell the Jones work in Chicago caught the attention of the *Baltimore Sun* late in May 1893. It was reported in the *Sun* that:

> *"Thomas A. Jones of Charles County has gone to Chicago to sell a book he has gotten up in connection with J. B. Mattingly, giving a history of the part Mr. Jones took in harboring Wilkes Booth after President Lincoln's assassination. They took with them* [on the train] *the sofa upon which Booth rested while his leg was being set and the bed he laid upon afterward and some other articles of furniture that were in the room* [of the Dr. Samuel A. Mudd house]

at the time." [53]

Information about the ultimate disposition of unsold copies of the book came to light about 1970. About 1917, John D. Rowe, then living on Mill Street (Kent Ave.) in La Plata, Maryland was hired by a neighbor to move all remaining copies from a corner of her barn, make a pile of them in front of the barn and burn them. And her will was done. While they were afire, John took one of them to his home and read it. He found that the books burning were the original copies of *J. Wilkes Booth*, which had been brought back from Chicago. John added his copy to the fire.

The Jones 1893 Chicago edition of *Booth*, in part because of its brevity and in part because it expresses Jones's long-held deep feelings about his commitment to J. W. Booth follows word for word starting on page 86.

In August 1893, Samuel Cox, Jr. made some marginal comments in a copy of the Jones book. These notes were added to a Charles County copy some years later through the courtesy of a grandson of Samuel Cox, Jr., who until recent years lived in New York State. Page numbers are those in the book itself.

(bottom of page 28)

An old cedar stump on the knoll where E. J. Collis's house now stands [on Wills Road] was used as a deposit for Confederate mail matter which was transferred by Col. Samuel Cox [to] Thomas A. Jones.

(page 60)

In 1877, after Dr. Samuel A. Mudd's return from Dry Tortugas and when he and myself were canvassing this county as the Democratic candidates for the Legislature, he told me he knew Booth but casually. That Booth had at one time sought an introduction to him through John H. Surratt on Pennsylvania Ave., in Washington. This was some time prior to the assassination, but he had refused and [said] that Booth had forced himself on him shortly afterward and that subsequently Booth attended church at Bryantown when he spoke to him but he was particular in not inviting him to his house, but that Booth came that evening unsolicited. He told me he was not favorably impressed with Booth and that when he, Booth and Harrold [sic], came to his home the night after the assassination they told him they were just from Virginia and that Booth's horse had fallen soon after leaving the river and had broken his leg, that he had rendered him medical assistance while in utter ignorance of the assassination.

---

[53] *Abstracts*, Volume 5: 1885-1893, p. 255.

J. WILKES BOOTH

AN HISTORIC, LITERARY MASTERPIECE

portraying the Author's connections with the Confederate mail, the Abduction Plot and Assassination of President Lincoln, Booth's dash through Southern Maryland, his crossing of the Potomac River, his death in Virginia and finally, the capture and imprisonment of the book's author gives the reader a heretofore untold chapter of that decade of American history.

J. WILKES BOOTH has been republished in its original form by The Society for the Restoration of Port Tobacco, Inc., and is bound in a beautiful hard back cloth cover with gold lettering, contains 126 pages and 18 original drawings. Its large type and lovely soft white pages are easy to read and will last a life-time.

J. WILKES BOOTH by Thomas A. Jones was first published in 1893. However there was such feeling in this Maryland area at that time that, it is reported, the entire publication was confiscated by local authorities and dumped into the Port Tobocco River.

Of that one and only edition there remain perhaps four or five copies now valued at $25.00. It was through the generousity of the owner of one of these rare books that the Society was able to republish this fine piece of literature and offer it to the American public at the amazingly low price of only $3.00

This bit of promotion copy was used by the Society for the Restoration of Port Tobacco about 1950 to announce their reprinting of the first edition of 1893.

(continued bottom of page 61)

That after he had set the broken leg, he, Dr. Mudd with letters he had but a short time gotten through the contraband mail for distribution, in going to Bryantown to mail them he was surprised to find the village surrounded by soldiers and upon being stopped by a sentry [there] he was horrified when told The President had been shot the night before. And, upon asking who had shot him the fellow had answered Booth. He then told me, his first impulse was to surrender Booth, that he had imposed upon him [and] had twice forced himself upon him and now the third time, had come with a lie upon his tongue and received medical assistance which would be certain to give him serious trouble, but he determined to go back and upbraid him for his treachery which he did. And [he said] that Booth had appealed to him in the name of his Mother whom he professed to love so devotedly and that he acted and spoke so tragically that he [Mudd] told them they must leave his house, which they then did and after getting in with Oswald Swann they were piloted to Rich Hill.

(bottom of page 68)

Booth and Harrold knocked at our door at about one o'clock Sunday morning. [Easter Sunday, 16th of April]

(bottom of page 71)

Harrold came on the front piazza [porch] and sounded the knocker. Booth sat on his horse in the yard under an alanthus [ailanthus] tree until Col. Cox opened the door. Harrold refused to give their names and Cox refused admittance. Booth approached the house to the piazza step and it was there by a brilliant moon Cox saw the letters J. W. B. on his arm.

(bottom of page 74)

The spot where Booth and Harrold were hiding is about half a mile south of the village of Cox's Station, on the line of the B&PRR [Baltimore and Potomac Railroad] and where Mary Tinker's house is now located, which was then a thicket of pines and on the lands owned at that time by Mrs. Eleanor Robertson, the original growth having been cut off...years before [The Collis-Mary Tinker house was built a few years after the Booth visit. Today this location is on Wills Road in Bel Alton and is the home of Mary Anne Burch, postmistress of the Bel Alton post office.]

(bottom of page 120)

Neither Booth nor Harrold ever entered Cox's house. I was up and saw them ride away in the morning light and pass out of the lot.

        S. Cox, Jr.

Inside the back cover, Samuel Cox, Jr., wrote:

"Ghost written by John Stone, a nephew of Judge Frederick Stone."
[Actually, John E. Stone, a La Plata lawyer-newspaper man, was a son of Jones's neighbor, Thomas David Stone, who was a first cousin of Judge Frederick Stone.]

       Thomas A. Jones's last years and breaths ended quite near where he began his turbulent, often unrewarding life. After his second marriage in Baltimore in the late 1870's to Margaret Rountree, Jones fathered his last child . . . son Edward A. born in 1880. Life in Baltimore was hard and certainly unglamourous. Most of the years here Jones managed a small feed, lumber and coal business. It was during this period that GATH established contact with him and that seemed to open new avenues for advantage and reward. We have no evidence that Jones profited in any grand way from Townsend's syndicated column presentations of the Booth-Herold-Jones episode in the great escape adventure disclosures of 1883-84. Doubtless GATH benefitted considerably from his research in Charles County and his productive friendship with Jones. GATH's diary reveals he paid Jones $60 for his contribution to the April 1884 *Century Magazine* article . . . not a great amount, but then the author received no fortune from the magazine for his work which surely fascinated Americans coast to coast. Townsend noted in pages of the December 1883 diary that the year of writing had been worth nearly $12,000 to him. This amount for such work marked GATH as a successful journalist in about the same league as Mark Twain.

For many years such roadside historic event signs have proclaimed to Charles County travelers the old story of the Booth-Jones saga. These two mark both the beginning and end of it from the curtain raiser at the Samuel Cox house through the closing excruciating measured steps over Mr. Dent's meadow to a steep descent to the little skiff in a sheltered pool at Potomac edge.

J. Wilkes Booth

An Account of His Sojourn in Southern Maryland after the Assassination of Abraham Lincoln, his Passage Across the Potomac, and his Death in Virginia

BY

Thomas A. Jones

The only living man who can tell the story

ILLUSTRATED

Copyright, 1893, by J. B. MATTINGLY

CHICAGO
LAIRD & LEE, PUBLISHERS
1893

TABLE OF CONTENTS

CHAPTER I.
	PAGE
INTRODUCTORY	7

CHAPTER II.
MY CONNECTION WITH THE CONFEDERATE MAIL 23

CHAPTER III.
THE ABDUCTION PLOT AND ASSASSINATION 39

CHAPTER IV.
BOOTH IS PLACED IN MY CHARGE 65

CHAPTER V.
I AM OFFERED $100,000.00 TO BETRAY BOOTH BY CAPTAIN WILLIAMS 83

CHAPTER VI.
I CONDUCT BOOTH TO THE POTOMAC 98

CHAPTER VII.
MY ARREST AND IMPRISONMENT 116

J. WILKES BOOTH.

CHAPTER I.

INTRODUCTORY

In writing this little book, it is my intention to tell the reader of the part I performed in the great war between the States, and my connection with the flight of the criminal whose deed closed the bloodiest chapter in our country's history.

No act ever committed has called forth such universal execration as the murder of that great and good man, Abraham Lincoln.

To-day I speak of the murdered President as "great and good," thirty years ago I regarded him only as the enemy of my country. But now that the waves of passion stirred up by the storm of war have all subsided and passed away forever, and I can form my opinions in the light of reason instead of the blindness of prejudice, I believe that Lincoln's name justly belongs among the first upon the deathless role of fame. I can now realize how truly he was beloved by the North, and what a cruel shock his death, coming when and as it did, must have been to the millions who held his name in reverence. And with that realization comes the wonder that the revenge taken for his murder stopped when it did.

I was born near Port Tobacco, Maryland, on October 2, 1820. Port Tobacco, the old courthouse village of Charles County, is situated at the head of Port Tobacco Creek, an inlet of the Potomac, that makes up to near the center of the county.

For forty years my life glided on as uneventfully as the lives of other farmers in Southern Maryland. I was married and had a large family. Born poor, by my habits of economy and industry I had worked my way up until I was in comfortable circumstances, and a competence was within my reach. But the war came on and I, like many of my neighbors, sacrificed my interest to the cause I believed to be just, and in the end shared in the general financial wreck of the South.

In 1861 I was living in my native county, on a farm of about five hundred acres which I had purchased a few years before. This farm was bounded on the west by the Potomac River, and on the north by Pope's Creek. My small, one-story frame house was built upon a bluff about eighty feet high. I could stand in my back yard and look up the river until my view was cut off by Maryland Point, seven or eight miles distant; while down the river I could see the water almost as far as the eye could reach.

My nearest neighbor was Major Roderick G. Watson. His land bounded mine on the south. He had a large, two-story high frame house, situated on a bluff at least one hundred feet above the water's level.

Abraham Lincoln

Pope's Creek, about sixty miles from Washington by water, though not more than forty by land, is a small stream emptying into the Potomac. It is nowhere more than forty or fifty feet wide; and is bordered on both sides by an extensive marsh. Between this marsh and the river there is a strip of land, about sixty feet wide, which forms a natural causeway between the two bluffs separated by the marsh. The public road approached the steamboat landing, then at this point, from two opposite directions. The boldness of the water and comparative narrowness of the river at Pope's Creek, together with its accessibility, made it a much used point of departure for those wishing to cross to Virginia. Besides, nearly every one in the neighborhood was known to be in sympathy with the South. It was, therefore, when the war had put an end to intercourse at Washington and above it with Virginia, that hundreds of people came to the neighborhood of Pope's Creek to get put across the river.

I entered with zeal into the Confederate cause. Scarcely a night passed that I did not take or send some one to Virginia. I have frequently crossed the river—which is not more than two miles wide from Pope's Creek —twice in a night, and sometimes oftener.

When it became known that a body of troops, under the command of General Sickles, was coming to Southern Maryland for the purpose of breaking up the contraband trade that had grown to such proportions as to have become a serious annoyance to the United States Government, there was a rush for Virginia by those on this side who had offended against the Government.

It was while this rush was going on that I left Pope's Creek one stormy evening about seven o'clock, with a boat load of ten or twelve men and women, bound for Virginia. I had in the boat two stout oarsmen, one of whom was a negro belonging to me named Henry Woodland—a man always faithful and true to me. I have not to-day a stauncher friend than old Henry.

Soon after leaving the creek the wind began to increase, and by the time we had got about one-third of the way across, it was blowing such a gale that I not only saw it would be impossible to proceed, but became considerably alarmed for our safety. Fortunately my passengers were ignorant of their danger and kept quiet. I watched for my opportunity, and just as the boat rose on a swell, gave the necessary command to the oarsmen and brought it safely around and put back to the Maryland shore. I took the interrupted travelers up to my house. On reaching there I learned that another party of four or five had been there during my absence looking for some one to put them across, and, finding I was not at home, had gone up to Major Watson's.

There were three schooners that had anchored in the cove opposite my house that evening, and I determined the best thing to be done under the circumstances was to capture one of them and use it in transferring the whole party across the river.

I walked over to Cliffton (Major Watson's) and explained to the party assembled there what I thought of doing, and asked which of the young men present would go with me and assist in making the capture. Several of them readily agreed to do so, and we set out for the creek. When we reached there we found another crowd, of a dozen or more, eager to get to Virginia. My party being joined by several volunteers, we got the rowboat out of the creek and boarded one of the vessels. The captain proved to be a timid man, and begged us not to force him to do what would be sure to get him into trouble with the Government. I told him I would see the captain of one of the other vessels, but if I did not succeed in making satisfactory arrangements, we would return and compel him to put us over.

The captain of the next vessel we boarded

was a man of altogether a different stamp. I think he would have "shown fight" had we not out-numbered his crew. After disputing the matter with us for a while and finding that we were positive, he gave in, remarking that if a man *had* to do a thing there was no use talking about it. We got the rest of our party aboard and made sail for the Virginia shore.

That was an unlucky night. No sooner were we well under way than the wind, that had been blowing such a gale that a small boat could not cross, died out, and it was sunrise before we succeeded in reaching the other side. Before landing we made up a purse for the captain, which he received with better grace than he had submitted to our coercion.

Among the many men whom I put over the river in those early days of the war was one Captain Emack.

Emack, while in Maryland on a recruiting expedition, was captured by a detachment of Union soldiers. The men who were taking him to Washington stopped for a while at a small village in Prince George's County, called T. B. Emack was left on the porch in charge of one of the soldiers; while the others entered the house. As soon as he found himself with only one man to guard him he determined to attempt his escape. He placed his hand, apparently, in his breast-pocket, and called the soldier as if he had something he wanted to show him. The unsuspecting soldier carelessly approached, but no sooner was he within striking distance than Emack drew a knife he had had concealed about his person and plunged it into him. The soldier fell and Emack ran for his life. Late that evening, foot-sore and weary, he arrived at the house of a gentleman, near Bryantown, who, he knew, was a warm sympathizer with the Confederacy. The gentleman of the house was absent, but the lady was at home, and to her Emack related his story, and appealed for aid. She took him in and ministered to his wants, and

after he had rested sent him on his journey in company with some other gentlemen who had stopped on their way to Virginia.

In September, '61, soon after the incidents just related, on returning home from one of my trips to Richmond, I was arrested and sent a prisoner to Washington. I was at first confined in a temporary prison which had been established on the corner of Fourteenth Street and Pennsylvania Avenue, but was soon removed to the old Capitol, and kept there six months.

During my imprisonment I made the acquaintance of several prominent and very interesting ladies and gentlemen who were fellow prisoners with me. Among them was Mrs. Greenhowe, of Washington, a widow lady, young and exceedingly handsome. She was tall and symmetrical in figure, with fine black eyes and dark hair. Her little daughter, Rosie, an attractive child about twelve years of age, was with her in the prison. After her release,

Mrs. Greenhowe was sent by the Confederate Government to England on some diplomatic mission, and on her return was drowned in the surf while attempting to land on the coast of North Carolina. Her daughter, who was with her, was saved.

I also met with two other widow ladies, agents in the Confederate service. One of them was a Mrs. Backsley, of Baltimore, an enthusiastic secessionist. I had visited her once, at her home in Baltimore, since the war. The other was Mrs. Morris, of Alexandria, Virginia.

Among the gentlemen of note who were in the Old Capitol with me were the Hon. Benjamin G. Harris, of St. Mary's County, ex-member of Congress from the 5th district of Maryland, and Captain George Thomas, of Alexandria, Virginia, a Confederate officer.

My old friend and neighbor, Mr. George Dent, and his son, were also fellow-prisoners with me.

I was released from the Old Capitol in March, 1862, by the general jail delivery ordered by congress at that time.

During my imprisonment, my old and esteemed friend, Major Watson, died. And soon after my return home my dear wife, whose health had been broken through care during my absence, was also taken from me. A bereaved and saddened man, I resumed my occupation of farming and fishing.

CHAPTER II.

MY CONNECTION WITH THE CONFEDERATE MAIL.

A short time prior to my arrest, among the many who came to my house to get put across the river was Major William Norris, of Baltimore County, Maryland, chief of the Confederate signal service.

He spent a night at my house, and in the morning walked out with me on the bluff that overlooks the river. He was struck with the extensive water-view from that point, and remarked to me: "What a place this would be for a signal station!"

Soon after my release from prison, Ben Grimes, of King George County, Virginia, whose house was just opposite mine, was sent to me, by Major Norris, to see if he could make arrangements with me to assist in carrying the Confederate mail from the United States and Canada to Richmond. At first I refused. I told Grimes that my duty to my children would not permit me to take the risk of imprisonment that such operations would involve. Grimes then represented to me that Major Norris had said that it was of the utmost importance to the Confederacy that it should have communication with points north of the Potomac, and that nowhere on the river was there a better location for a signal station than the bluffs near Pope's Creek, or a more suitable place for putting the mail across the river than off my shore.

After thinking the matter carefully over I agreed that, if I was given the entire control of the ferry and all the agents in Maryland, and also allowed a voice in the management on the other side of the river, I would undertake the work.

Grimes informed Major Norris of what I had

said, and the Major then held an interview with me. He seemed to think that I possessed the requisite qualifications for the work in hand and readily agreed to the proposition I had made to him through Grimes.

It required great caution and unrelaxing vigilance to successfully carry on the operations in which I was now engaged. The river was filled with gunboats plying up and down, day and night. An armed patrol guarded the shore and the Federal Government had a spy upon nearly every river farm in Southern Maryland. There was a detachment of troops stationed at Pope's Creek, and another on Major Watson's place—not three hundred yards from my house. Captain Groff was in command of these troops, and had his headquarters at Ledlow's Ferry, the next place below Major Watson's. Captain Groff was succeeded in command of these men by Captains Boyle and Watkins.

The signal camp on the Virginia side was established in the low, swampy grounds back of Grimes'. It was under the command of Lieutenant Charles Caywood. The boats used in the mail service were kept on the Virginia side.

I had noticed that a little before sunset the reflection of the high bluffs near Pope's Creek extended out into the Potomac till it nearly met the shadow cast by the Virginia woods, and therefore, at that time of evening it was very difficult to observe as small an object floating on the river as a rowboat. The pickets did not go on duty until after sunset. It was therefore arranged that if the coast was clear the boat from Grimes' should come across just before sunset, deposit the packets from Richmond in the fork of a dead tree lying on my shore, and take back the packet from the North found there. Unless for some especial reason, I would not be on the beach when the boat arrived.

If it was not safe for the boat to cross, a

black signal was hung in a certain one of the high dormer-windows of Major Watson's house. The person who attended to this signal was Miss Mary Watson.

Miss Watson was a remarkably pretty young lady, twenty-three or twenty-four years of age. She had a mass of black hair, dark eyes shaded by long lashes that made them appear even darker, and heavy black brows. Her carriage was erect, and figure slender, which made her appear a little above the average height.

She loved the Confederacy with an ardor so intense that I believe, for its sake, she would have made almost any sacrifice. I know that I owe, in a great measure, the successful management of the Confederate mail to her ceaseless vigilance and untiring zeal.

About the close of the war she married a Dr. Carvell, a blockade-runner, and went with him to California.

I do not know whether or not she will ever read these lines; but if she should, I would have

her know that my old heart grows warm, and my dim eyes dimmer when I think of her in her youth and beauty tirelessly laboring by my side in the cause we both fondly loved.

The pickets went off duty in the morning. Sometime during the day I would go down to the shore and get the packet left there the evening before. Letters, etc., going north, were addressed to the parties for whom they were intended. All I would have to do with them was to put them in a postoffice. I seldom posted them at my nearest office, which was Allen's Fresh, about three miles distant, for fear of exciting suspicion, but would send them by our trusted agents to be posted at different places some distance off.

Especially important matter was never sent by the mail, but was always intrusted to our agents.

Stowten W. Dent, M. D., and my brother-in-law, Thomas H. Harbin, were two of our most active agents.

Dr. Dent, who died in 1883 at the advanced age of eighty, had two sons in the Confederate army. His home was near Centerville, a small village not far from the central part of Charles County. He was a practicing physician, and used to make his professional rounds on horseback. In winter he invariably wore an overcoat that came down below his knees, provided with numerous and capacious pockets, and high boots; and in summer, a long linen duster also well provided with pockets. The number of letters and papers he could conceal in his pockets and boot legs was astonishing. Some one in the neighborhood of Pope's Creek was always sick. Scarcely a day passed that some member of the Watson family or nine did not need Dr. Dent. He came and went unquestioned and unsuspected. He would take the mail as far as Port Tobacco, ten miles from my home; or Bryantown, a village in the eastern part of the county about fifteen miles from Pope's Creek; or even as far as Charlotte Hall in St. Mary's County, fully twenty miles off, and then transfer it to some other agent who would convey it further on toward its destination.

The doctor had a son named Warren, a mere lad of about ten years of age, who, child though he was, was as energetic, discreet and intelligent as any agent in the Confederate service. The most important matter was often intrusted to his care, and always safely intrusted.

A Canadian by the name of Williams, another Confederate agent, was frequently in the neighborhood of Pope's Creek, waiting to see some man who never came. He would take the mail for Canada clear through.

Mail matter going South when it came by the United States mail-route, was always addressed to me. I would get the Northern newspapers into Richmond within twenty-four hours after their publication. But a much larger portion of the mail going South came to me through the hands of our agents.

Quite a prominent gentleman of Prince George's County, Maryland, was very active as a Confederate mail-agent. He turned his attention to gardening, and, as he lived but a short distance from Washington City, would drive his wagon into town and get a load of manure, in which he would hide the matter destined for the South, and bring it safely out.

Every packet, going north or south, was conveyed to and from the boat in my own hands. I trusted no one that it was not absolutely necessary to trust. The result was that not in a single instance was I betrayed.

Sometimes I would give some mail-matter to Miss Watson to be taken care of until it could be forwarded, but generally I kept it myself in a small, dark closet up stairs in my house. This closet had a door just large enough to admit a man's head and shoulders. It was necessary to stoop down to see into it and with your head in the door you would have to turn

over on your back to be able to see directly over your head; and right there was where I hid the Confederate mail. And though the house was frequently searched while the mail was in it, nothing was ever discovered.

The United States Government knew, of course, that communication was going on between the South and North of the Potomac; and it exhausted its ingenuity in vain in trying to discover how it was managed. This failure to discover our method gave rise to some far-fetched conjectures as to how we managed, despite the vigilance of the Federal Government, to escape detection. One idea advanced was that by some ingenious contrivance the mail was drawn from one side of the river to the other, *under the water*.

From the time I accepted the position of chief signal agent north of the Potomac, which was in the spring of '62, till the close of the war, there was scarcely an evening that the boat did not make its trip across the river; *and not one letter or paper was ever lost.*

The only thing that miscarried the three years of my service was some baggage belonging to Mrs. William N. Norris. It should never have been attempted to send this baggage by our mail-route. It was contrary to my instructions that anything not strictly mail-matter should ever be sent by our agents.

Mrs. Norris, who was living in Baltimore, wanted to join her husband in Richmond. So Lieutenant Carey was sent over to Pope's Creek one evening in our mail boat to take her across the river. The cart containing her baggage was, for some reason, late in reaching the landing and it was determined that instead of waiting for it, which would have been risky, Lieutenant Carey should take Mrs. Norris across and return next evening for her baggage. The next evening, just as the boat touched the shore at its usual time, the vehicle bringing the baggage was seen coming down the hill to the landing. But, near at hand as it was, Fate had decreed that Mrs. Norris' effects should never cross the river. A gunboat, which none of us had observed through the gloom of evening, ran into the cove a little below our landing and sent out a small boat to patrol along the shore. Just as the cart bringing the baggage was nearing us, I heard the dash of oars and caught a glimpse of the approaching boat. "Take care of yourselves boys," I cried, and ran for my horse that was tied a short distance down the beach. The two men who were with Carey were well acquainted with the country around, and made their escape without trouble. But Carey, who had never been at Pope's Creek before, ran the wrong way and got into the marsh. Our boat and the baggage were captured but we saved the mail.

Late that evening I was disturbed by a knock at my door and on going out found it was Carey who had aroused me. He had escaped capture but at the expense of getting into one of the muskrat-leads with which the marsh was honeycombed. He presented a woeful appearance, being literally covered with black marsh-mud. I was sorry for him but could do nothing toward getting him across the river that night, as the gunboat was still in the neighborhood, and the soldiers were on the alert. I was afraid even to take him in, so I pointed out a piece of pines back of my barn where he could pass the night, and promised to help him next day when the pickets went off duty.

In passing my barn on his way to the pines that night, Carey picked up a turn of straw and took it along with him to serve as bedding. I traced him to his hiding-place, next morning, by the straws he had dropped upon his way. I found his appearance by no means improved by his night in the woods; and the partially dried mud with which he was covered looked blacker than it had seemed by candle light. He had taken his wet shoes and stockings off to warm his feet, he said, in the sunshine, and

I remember well to this day how cold and red his feet looked. I took care of him that day and at night he got back to Virginia.

In April, 1865, I went to Richmond to collect what was due me by the Confederate Government. I had to remain a few days and so happened to be there when the Southern Army evacuated the city.

When the Confederacy expired, not only did I lose the two thousand and three hundred dollars due me for my three years' service, but also three thousand dollars that, early in the war, I had invested in Confederate bonds.

About the close of the war I sold my place at Pope's Creek and removed with my family to a farm called Huckleberry, about two miles north of my former residence. Huckleberry house is situated about one hundred yards back from the public road and one mile from the Potomac river.

CHAPTER III.

THE ABDUCTION PLOT AND ASSASSINATION.

Sometime in December, 1864, I heard that there was "a big scheme" afoot to abduct President Lincoln and take him a prisoner to Washington.

Briefly stated, the plan was this: The President, when he went for his customary evening drive toward the Navy Yard, was to be seized and either chloroformed or gagged, and driven quietly out of the city. If in crossing the Navy Yard bridge the carriage should be stopped, the captors would point to the President and drive on. The carriage was to be escorted out of the city by men dressed in Federal uniform. Relays of fast horses were in readiness all along the route, and a boat in

"Huckleberry," Residence of Thomas A. Jones

Mrs. Surratt's House at Washington

J. WILKES BOOTH

which to take the captive across the Potomac was kept on the west side of Port Tobacco Creek, about three and a half miles from the town of the same name.

There was not much danger that the carriage containing the abducted President, once clear of the city, would be overtaken, as it would be impossible for the pursuers to obtain fresh horses. The distance to be traversed to reach the Potomac was only about thirty miles, and with the boat and men to row it in readiness, the river could be crossed within an hour of the time it was reached.

The idea of the conspirators was that with such a hostage in its power the Confederacy would be able to dictate terms to the North.

Had it not been for the wretched condition of the roads during the latter part of the winter and early spring, due to the mild weather, frequent rains and constant hauling over them of the heavy army wagons, there is no doubt but what the execution of this plan would have

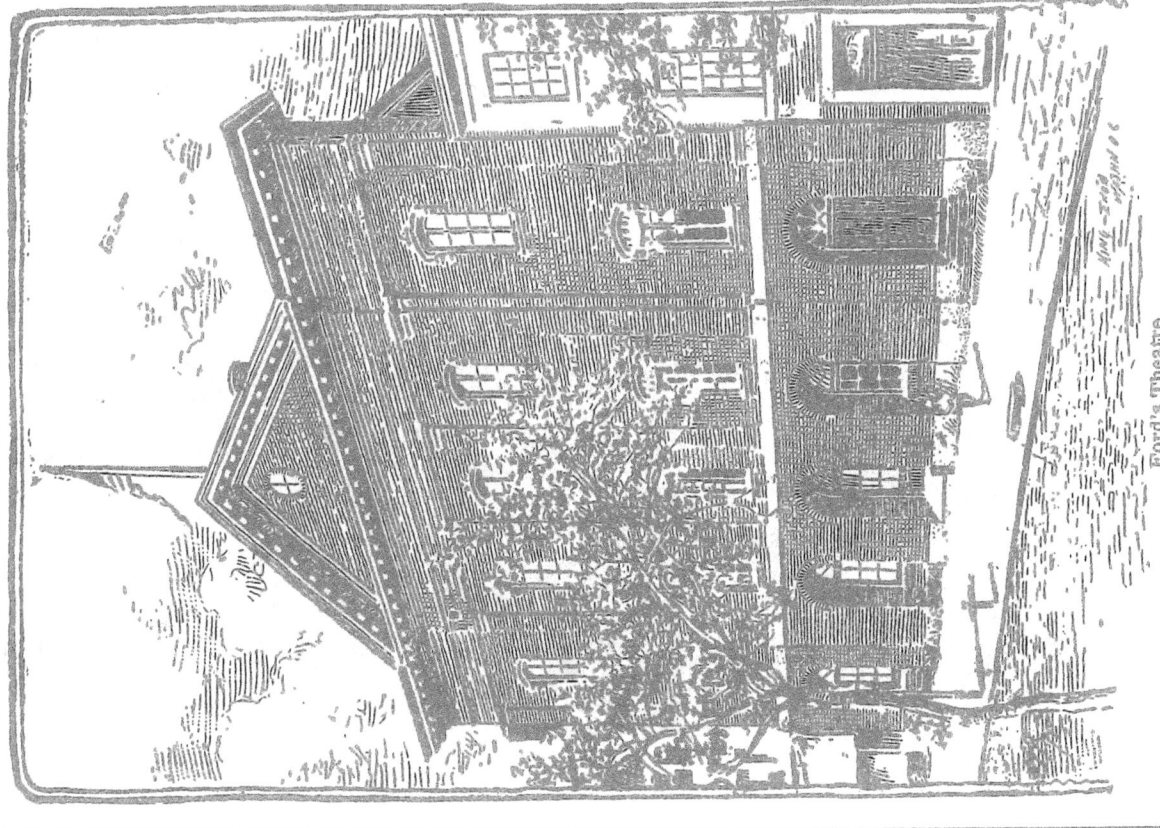
Ford's Theatre

J. WILKES BOOTH

been attempted; and when we consider the thoroughness and secrecy with which the plotters made their arrangements, it does not seem improbable that it might have been successfully accomplished.

There were quite a number of persons in this abduction conspiracy; prominent among whom were the actor, John Wilkes Booth, and his friend, John H. Surratt.

The house of Mrs. Surratt,—mother of John H. Surratt,—in Washington City, was the frequent *rendezvous* of the daring conspirators. The arms and ammunition that would be needed in carrying out the contemplated enterprise were placed at Surratt's house, Surrattsville, Prince George's County, Maryland, eight miles south-east of Washington.

I do not think the abduction plan was ever given up until Booth killed his victim instead of capturing him.

It is generally believed that the assassin did not determine upon the murder of the Presi-

The Assassination of President Lincoln

---

dent until the morning of the day he committed the dreadful deed. Be that as it may, when the evening of that fatal Good Friday arrived, the plans for the intended quadruple murder were all arranged. Payne and Atzerodt, acting under Booth's instructions, were to dispatch Secretaries Seward and Stanton, while Booth himself undertook to kill the President and General Grant, who was expected to accompany him to Ford's theater.

After the murder was committed, Booth's intention was, to make his way to the Potomac, guided by David E. Herold, who professed to be familiar with the roads through Southern Maryland, and cross the river in the boat that had been placed by the abduction plotters at Port Tobacco Creek, months before.

I shall not attempt to give a detailed account of the assassination. To do so would be but to repeat what has often been told before.

The attempt upon Secretary Stanton, for some reason, was not made. Payne entered Mr.

Seward's sick room and made a violent and well nigh fatal attack upon him, but fortunately did not succeed in killing him. General Grant did not go with Mr. Lincoln to the theater that night and thereby escaped.

How Booth entered the box behind the President, fired the fatal shot, stabbed Major Rathbone, who attempted to interrupt him, leaped from the box, and catching his spur in the drapery, fell, and fractured his leg; sprang upon the stage and waving his bloody knife, exclaimed, "*Sic semper tyrannis*," and then sped away through the darkness—has become history, and I need not dwell over it.

The dying President was removed from the theater to that house on 10th Street, which since that memorable night has become an object of national interest. He lingered in an unconscious condition for several hours and then—Secretary Stanton's voice broke silence: "And now he belongs to the ages."

Booth made his escape from the rear of the theater immediately upon 9th Street, thence to Pennsylvania Avenue. When he reached the avenue he rode down toward the Capitol as far as 11th Street, then down 11th Street to the Navy Yard bridge, where he was joined by Herold. They crossed the bridge unchallenged by the guards and took the road toward Surrattsville. When they reached that village they paused a moment to get the carbines and ammunition which, as has been before stated, were placed there during the arrangement of the abduction plot. Having obtained these articles—they were handed out to them by a man named Lloyd—the fugitives continued their flight.

By this time Booth's leg had become so painful that he abandoned the original intention of taking the shorter route to the river, and crossing in the boat that had been placed at Port Tobacco; instead, he took the road that bears farther toward the east, passed through T. B., a small village about five miles from

The House in which Lincoln Died

Death of President Lincoln

Surrattsville, without stopping, and arrived at Dr. Mudd's some time in the early morning. Dr. Samuel A. Mudd lived in the northeastern part of Charles County, about twenty miles from Washington, and not far from the little village of Bryantown. Booth knew the doctor, having met him and visited at his house when in the county about eighteen months before the assassination. The statement he made to the doctor was that his horse had fallen and hurt him. Both he and Herold entered the house and the doctor, assisted by his kind-hearted wife, who had arisen for the purpose, proceeded to examine and dress the fracture. The fugitives remained there until the following evening. How much they told Dr. Mudd beyond the fact that they wished to cross the river to Virginia, is not known.

Booth kept his bed next morning and the meals that were sent up to him were returned untouched. Finding he ate nothing, Mrs. Mudd went up to his room and inquired if there was anything he would like. He asked if she had any brandy. She told him no, but that she had in the house some good whiskey and offered him a glass of that, but he declined it.

While at breakfast, Herold made inquiries concerning the different people in the neighborhood and the roads to the river. He inquired very particularly as to the location of the residence of the Rev. Lemuel Wilmer, rector of Port Tobacco parish, a gentleman well known, far and near, as a staunch and loyal Unionist.

After breakfast Herold expressed a desire that he and his companion should resume their journey. Dr. Mudd told him that the crippled man was in no condition to travel on horseback and suggested that he (Herold) should ride over to a certain neighbor's and borrow a carriage in which to take his friend on. Herold acted upon this suggestion, but the gentleman to whom he applied told him that as next day would be Easter Sunday his family was anxious

Residence of Dr. Samuel A. Mudd

Surratt's House at Surrattsville

to attend church, and therefore he could not spare the vehicle.

While Herold was gone to look for a carriage, Doctor Mudd rode out toward Bryantown, to visit some patients. On his way he met his cousin, George Mudd, who told him there were soldiers in Bryantown who had brought the astounding news from Washington that President Lincoln had been killed, and that the assassin had taken the road toward Southern Maryland. The doctor was much disturbed by what he heard, and told his cousin of the two visitors at his house. When he parted with his cousin, instead of going on to Bryantown he returned home.

In the meantime Herold had returned from his fruitless search for a carriage and after having an interview with Booth, assisted him down the stairs and not heeding the remonstrances of Mrs. Mudd, whose feelings of hospitality could ill brook that a guest unfit to travel should leave her house, they mounted their horses and rode away.

Booth carried with him from Dr. Mudd's a rude crutch which a colored man on the place had hastily made for him.

The doctor doubtless felt much relieved when, on reaching home, he found his dangerous visitors had gone.

When the fugitives left Dr. Mudd's late that Saturday evening they did not go through Bryantown, but crossed Zachiah swamp higher up. They lost their way in the swamp and rode back and forth, as the tracks of their horses showed. It must have been while they were trying to find the direct road out of the swamp that they came across the negro, Oswald Swann, who guided them to Cox's.

CHAPTER IV.

BOOTH IS PLACED IN MY CHARGE.

Saturday evening, just about the time Booth and Herold were setting out from Dr. Mudd's, I was over at my former residence, near Pope's Creek, attending to some business, when two Federal soldiers rode up and asked me whose boat that was down in the creek. I told them it was mine. "Well," one of them replied, "you had better keep an eye to it. There are suspicious characters somewhere in the neighborhood who will be wanting to cross the river, and if you don't look sharp you will lose your boat."

"Indeed," I answered, "I will look after it. I would not like to lose it, as it is my fishing boat and the shad are beginning to run."

Room at Doctor Mudd's where Booth Slept; Left-hand Bed the one Booth Slept in

The two soldiers then conversed in an undertone with each other for a few moments when the one who had first spoken turned to me and said: "Have you heard the news friend?"

I answered "No."

"Then I will tell you," said he. "Our President was assassinated at ten o'clock last night."

"Is it possible!" I exclaimed.

"Yes," he returned, "and the men who did it came this way." They then rode off and I soon afterward returned home.

The next morning, which was Easter Sunday, soon after breakfast, Samuel Cox, Jr., adopted son of my foster-brother, Samuel Cox, came to my house, Huckleberry, and told me his father wanted to see me about getting some seed-corn from me. He added, in an undertone, "Some strangers were at our house last night."

Even had I not heard the evening before of the assassination of Mr. Lincoln, knowing Cox as I did, I would have been sure he had sent for me to come to him for something of more importance than to talk about the purchase of seed-corn. But putting together the intelligence I had the evening before received from the two soldiers, the fact that strangers had been at Cox's the previous night and that Cox had now sent for me, I was convinced that he wanted to see me in reference to something connected, in some way, with the assassination.

I had my horse saddled and young Cox and I rode together to Cox's home, Rich Hill, about four miles to the north-east of my house. Though twenty-eight years ago, I remember well my ride that bright spring morning.

My companion and I spoke but little, and that little upon indifferent matters. My mind was too much occupied in speculating upon what was to be the outcome of my approaching interview with Cox, to allow me to enter

Residence of Samuel Cox

into conversation unless upon the subject then absorbing my attention; and my four years varied experience during the war, as well as my innate prudence, forbade me to talk, when not absolutely necessary, upon subjects that might be dangerous. Therefore, with the exception of a casual remark or two as to the roads or the weather, we rode on in silence.

It was a little after nine o'clock when we reached Rich Hill. Cox met me at the gate and we walked off a short distance to an open space where there was nothing that might conceal a listener.

I have often observed when there is a weighty matter to be discussed between men, how reluctant they seem to approach it. Cox had a most important disclosure to make to me; I knew that he had, and yet, for some minutes, we spoke of any matter rather than that which had brought us together. At length he said to me: "Tom, I had visitors about four o'clock this morning."

"Who were they, and what did they want?" I asked.

"They want to get across the river," said Cox, answering my last question first; and then added in a whisper, "Have you heard that Lincoln was killed Friday night?"

I said, "Yes, I have heard it," and then told him of my interview with the two soldiers the evening before. When I had finished there was silence between us for a minute, which was broken by Cox.

"Tom, we must get those men who were here this morning across the river."

He then went on to say that about four o'clock that morning he was disturbed by a knocking at the door. On opening the door he found a strange man standing there, while waiting at the gate was another stranger on horseback accompanied by a negro man of the neighborhood, named Oswald Swann. He went out to the man on horseback who, upon being satisfied that it was Samuel Cox he ad-

dressed, took him a little apart, out of hearing of the negro, and told him what he had done. He showed him in India ink upon his wrist the initials of J. W. B. He told Cox that he knew he (Cox) was a Southern sympathizer who had worked for the Confederacy and that he threw himself upon his mercy. He explained how he had broken his leg and thereby been prevented from reaching the river that fateful Friday night. All this Cox told me, speaking almost in a whisper, standing near his gate that bright Easter morn.

After he had finished his recital he said to me again, "Tom, you must get him across."

Reader, it will scarcely surprise you when I say I was much disturbed by Cox's disclosure.

In the cause of the Confederacy I was willing to risk my life as I had done often. But the war was over. The cause which I loved and for which I had labored was lost. Nothing now could raise from the dust the trailing Stars and Bars. I knew that to assist in any

J. WILKES BOOTH 73

way the assassin of Mr. Lincoln would be to put my life in jeopardy. I knew that the whole of southern Maryland would soon be—nay, was even then—swarming with soldiers and detectives, like bloodhounds on the trail, eager to avenge the murder of their beloved President and reap their reward. I hesitated for a moment as I weighed these matters.

I was aroused by Cox's voice:

"Tom, can't you put these men across?"

"Sam," I replied, "I will see what I can do, but the odds are against me. I must see these men; where are they?"

He then told me that he had sent them to a place in a thick piece of pine about one mile to the west of his house—his overseer, Franklin Robey, guided them to the spot, with the promise that he would send some one to them; and had advised them to remain perfectly quiet. He agreed with them upon a signal by which they might know the man who came to them was from him. This signal was a

74 J. WILKES BOOTH

peculiar whistle which I do not remember now. He also provided them with sufficient food for the day, and I believe, though I am not sure, with a pair of blankets.

"Take care how you approach them, Tom," Cox said to me as I was leaving. "They are fully armed and might shoot you through mistake."

I left Cox and rode toward the spot he had indicated.

It was with extreme reluctance I entered upon this hazardous enterprise. But I did not hesitate; my word was passed.

The place where Booth and Herold were in hiding was about two hundred yards south of the present village of Cox Station, which is five miles from Pope's Creek, the southern terminus of the Baltimore and Potomac Railroad. An Englishman, named Collis, now occupies a house built upon the exact spot where I first beheld the fugitives.

As I drew near the hiding place I saw a bay

mare, with saddle and bridle on, grazing in a small open space where a clearing had been made for a tobacco bed. I at first thought that she belonged to some one in the neighborhood and had got away. I caught her and tied her to a tree. I then went on a little further until I thought I was near the place indicated by Cox. I stopped and gave the whistle. Presently a young man—he looked scarcely more than a boy —came cautiously out of the thicket and stood before me. He carried a carbine ready cocked in his hands.

"Who are you, and what do you want?" he demanded.

"I come from Cox," I replied; "he told me I would find you here. I am a friend; you have nothing to fear from me."

He looked searchingly at me for a moment and then said, "Follow me," and led the way for about thirty yards into the thick undergrowth to where his companion was lying. "This friend comes from Captain Cox," he

Collis's House

said; and that was my introduction to John Wilkes Booth. He was lying on the ground with his head supported on his hand. His carbine, pistols and knife were close beside him. A blanket was drawn partly over him. His slouch hat and crutch were lying by him. He was dressed in dark—I think black—clothes; and though they were travel-stained, his appearance was respectable. Booth has been so often described that I will not attempt a description further than to say that, though he was exceedingly pale and his features bore the evident traces of suffering, I have seldom, if ever, seen a more strikingly handsome man. He wore a mustache and his beard had been trimmed about two or three days before. His voice was pleasant; and though he seemed to be suffering intense pain from his broken leg, his manner was courteous and polite.

I said that I had entered with reluctance upon the dangerous task of succoring Booth. But no sooner had I seen him in his helpless and suffering condition than I gave my whole mind to the problem of how to get him across the river. Murderer though I knew him to be, his condition so enlisted my sympathy in his behalf that my horror of his deed was almost forgotten in my compassion for the man, and I felt it my bounden duty to do all I could to aid him; and I made up my mind, be the consequences to me what they might, from that time forth my every energy should be bent to the accomplishment of what then seemed to be the well-nigh hopeless task of getting him to Virginia.

I told him that I would do what I could to help him; but for the present he must remain where he was; that it would not do to stir during the hue and cry then being made in the neighborhood. I promised to bring him food every day, and to get him across the river, if possible, just as soon as it would not be suicidal to make the attempt.

He held out his hand and thanked me.

He told me, as he had told Cox, that he had killed President Lincoln. He said he knew the United States Government would use every means in its power to secure his capture. "But," he added, with a flash of determination lighting up his dark eyes, "John Wilkes Booth will never be taken alive;" and as I looked at him, I believed him.

He seemed very desirous to know what the world thought of his deed, and asked me to bring him some newspapers.

I mentioned to Booth that I had seen a horse grazing near by, and he said it belonged to him. I told him and Herold that they would have to get rid of their horses or they would certainly betray them; besides, it would be impossible to feed them.

Before leaving, I pointed out to Herold a spring about thirty or forty yards distant, where he could procure water for himself and companion. I advised him to be very cautious in going to the spring, as there was a footpath running near it that was sometimes, though seldom, used. Then promising to see them next day and bring food and newspapers, I mounted my horse and rode home.

I may as well insert here the sequel to my advice concerning Booth's and Herold's horses.

During my week's attendance on the two men I never once saw Herold's horse, and saw Booth's only on the one occasion already referred to. I had no hand in the disposition of them; and do not remember, if I ever knew, the exact day that Herold removed them.

After the fugitives crossed the river, and just before I was arrested, Cox told me that he stood on the hill near his house and saw Herold taking the two horses down toward Zachiah Swamp and heard the two reports of the pistol that killed them.

It has been stated that Herold buried the horses after he shot them. I am sure that is a mistake. To have done so, he would have required assistance; and, besides, newly dug earth would scarcely have escaped detection during the scrutinizing search made from one end of Zachiah Swamp to the other.

Cox said that sometime after the horses were killed he rode down to the spot where he had heard the pistol discharged and searched minutely in every thicket or brier-clump in the neighborhood, but could not discover a trace of them.

In the dense growth that covers the swamp there is a large area of quicksand covered with water. It is my opinion that the horses were led into this quick-sand and shot there, and that their own weight sunk them.

Whether my opinion is correct or not, it is certain that not even a bone of them has ever been discovered to this day.

CHAPTER V.

I AM OFFERED $100,000 TO BETRAY BOOTH, BY CAPTAIN WILLIAMS

As I rode homeward that Sunday morning after my interview with Booth and thought over all that had just transpired, I realized the danger of my situation and the grave responsibility I had assumed.

I knew that what I was then engaged in, even if known, would be regarded in those heated days of bitter sectional hatred by the one side, as just and right; by the other as darkly criminal. I felt confident that the people of the South, whatever might be their public expressions on the subject, would not blame me whether I succeeded or failed; while in the North I would be looked upon as the vile aider and abetor of a wretch stained with as dark a crime as the recording angel ever wrote down in the eternal book of doom.

The calm and dispassionate judging people of to-day, both North and South, will regard what I did—if they deign to consider at all the acts of so humble an individual—as deserving neither praise nor blame, but as being the natural result of the times and circumstances.

There were but two boats on this side the river that I knew of, and they were both mine. One was the little bateau in Pope's Creek already referred to; the other, a somewhat larger boat, also in the creek, but higher up the stream, and hidden in the marsh grass.

I felt certain that the bateau which had been spoken of by the two soldiers on Saturday would either be watched by them or secured. And why the Government employees did not take possession of that boat at once has always been a mystery to me.

It need not be said that Booth's only chance for crossing the river depended upon my being able to retain possession and control of one of these two boats.

When I reached home from my visit to Booth that Sunday, I called Henry Woodland, who had continued to live with me after his emancipation, and told him to get out some gill-nets next morning and to fish them regularly every day, and after fishing always to return the boat to Dent's Meadow.

Dent's Meadow was then a very retired spot back of Huckleberry farm, about one and a half miles north of Pope's Creek, at least a mile from the public road and with no dwelling house in sight. This meadow is a narrow valley opening to the river between high and steep cliffs that were then heavily timbered and covered with an almost impenetrable undergrowth of laurel. A small stream flows through the meadow, widening into a little creek as it approaches the river. It was from this spot I determined to make the attempt of sending Booth across to Virginia.

Dent's Meadow

Immediately after breakfast on Monday morning, I wrapped up some bread and butter and ham, filled a flask with coffee, and put it all in the pockets of my overcoat. I then took a basket of corn on my arm as though I was going to call my hogs that ran at large in the woods surrounding my house, and mounting my horse, set out on my dangerous visit to the daring assassin.

I rode up the public road in a northerly direction, till I reached a place, about a mile distant from my house, where a cart-track branched off toward the northeast. I turned into this path and rode leisurely on till I was within a hundred yards of the hiding-place. I then dismounted, led my horse into a thicket and tied him, and then went forward and gave the signal.

Nothing of any especial importance happened at this interview. Booth seemed to be suffering more with his leg than on the previous day, and was impatient to resume his journey so as to reach some place where he could be housed and get medical attention. I told him he must wait. While we were talking I heard the clanking of sabers and tramping of horses, as a body of cavalry passed down the road within two hundred yards of us. We listened with suspended breath until the sound died away in the distance. I then said, "You see, my friend, we must wait."

"Yes," he answered, "I leave it all with you."

I left him and took my way to Cox's.

I found Cox anxious to see me and learn what I had done and what I intended to do. I explained to him, as I had done to Booth, my plans. Briefly stated, they were simply to keep myself informed as to what was going on; and the first night the neighborhood was clear of soldiers and detectives to get my charges to the river at Dent's Meadow and let them cross in the boat I kept there for that purpose.

Tuesday morning, after my visit to the pine thicket, I rode up to Port Tobacco.

Tuesday was then, as it is now, the day for the transaction of public business in our county. I was therefore likely to meet a good many people in the county-town that day, and hear whatever was going on.

I found the men gathered about in little groups on the square, as men in villages will always be found when anything of more than usual interest is engaging public attention. Upon this occasion, of course, they were discussing the assassination, and the probable whereabouts of the assassin. The general impression seemed to be that Booth had not crossed the river.

I mingled with the people and listened till I was satisfied that nothing was positively known. Every expression was merely surmise.

It was while in Port Tobacco that day I made the acquaintance of Captain Williams. He was standing in the bar-room of the old Brawner Hotel (now St. Charles Hotel) in the act of drinking with several gentlemen who were gathered around him, when I entered. Some one introduced me to him and he politely invited me to drink with him. Just as we were about to take the drink, standing with our glasses in our hands, he turned to me and said, "I will give one hundred thousand dollars to any one who will give me the information that will lead to Booth's capture."

I replied, "That is a large sum of money and ought to get him if money can do it."

In Mr. George Alfred Townsend's article, "How Wilkes Booth Crossed the Potomac," published in the Century magazine of April, 1884, the author comments upon this offer made in my presence and partly to me, in the following terms: "When we consider that the end of the war had come and all the Confederate hopes were blasted and every man's slave set free, we may reflect upon the fidelity of this poor man whose land was not his own and with inevitable poverty before him perhaps for the

Brawner's Hotel, Port Tobacco

rest of his days," etc. It appears from this that Mr. Townsend thinks I deserve some meed of praise for not being bribed to betray what I considered a sacred trust. But it seems to me that, had I, *for money*, betrayed the man whose hand I had taken, whose confidence I had won, and to whom I had promised succor, I would have been, of all traitors, the most abject and despicable. Money won by such vile means would have been accursed and the pale face of the man whose life I had sold, would have haunted me to my grave. True, the hopes of the Confederacy *were* like autumn leaves when the blast has swept by. True, the little I had accumulated through twenty years of unremitting toil *was* irrevocably lost. But, thank God, there was something I still possessed—something I still could call my own, and its name was Honor.

In 1889, soon after I was dismissed from the humble position I had held under the Federal Government in the Navy Yard at Washington, I met, for the first time since those memorable and eventful days of which I have been writing, Captain Williams. He was then a detective in Washington City. In the interview I then had with him (a not very accurate account of which was published in the newspapers at the time) Captain Williams told me that that day in Port Tobacco he very strongly suspected I knew more than I was willing to tell. But there was certainly nothing in his manner from which I could have inferred that he was any more suspicious of me than he was of any one else in southern Maryland.

But to return to my story.

Wednesday and Thursday passed uneventfully away. The neighborhood was filled with cavalrymen and detectives. They visited my house several times during that week (as they did every house in southern Maryland) and upon one occasion searched it. They also interviewed my colored man, Henry Woodland,

and threatened him with dire penalties if he did not tell all he knew. Henry did not *know* anything because I had told him nothing. I took no one into my confidence. But he could not but suspect that something unusual was going on. But I was not uneasy on that score, as I had thorough reliance in his fidelity to myself, and his discretion.

As the days rolled away, Booth's impatience to cross the river became almost insufferable. His leg, from neglect and exposure, had become terribly swollen and inflamed, and the pain he had to bear was excruciating. To add to his further discomfiture—if that was possible—a cold, cloudy, damp spell of weather, such as we often have in spring, set in and continued throughout the week. Fortunately, though, there was no rain. Trying as was the situation, it had to be endured. The time to move had not yet arrived. So through six long, wearisome days, and five dark and restless nights, Booth lay there in hiding. The only breaks in the monotony of that week were my daily visits, and the food and newspapers I carried him. He never tired of the newspapers. And there—surrounded by the sighing pines, he read the world's just condemnation of his deed and the price that was offered for his life.

CHAPTER VI.

I CONDUCT BOOTH TO THE POTOMAC.

On Friday evening, one week after the assassination, I rode down to Allen's Fresh. I think I had been there every day Booth had been under my care, except the Tuesday I went to Port Tobacco.

Allen's Fresh, about three miles east of my house, was and still is, a small village situated where Zachiah Swamp ends and the Wicomico River begins.

I had not been long in the village when a body of cavalry, guided by a man from St. Mary's County named John R. Walton, rode in and dismounted. Some of the soldiers entered Colton's store, where I was sitting, and called for something to drink. Soon afterward Walton came in and exclaimed, "Boys, I have news that they have been seen in St. Mary's," whereupon they all hastily remounted their horses and galloped off across the bridge in the direction of St. Mary's County.

I was confident there were no other soldiers in the neighborhood.

"Now or never," I thought, "is my chance."

I waited a few minutes so as not to excite suspicion by leaving immediately after the soldiers, then mounted my horse and rode slowly out of the village.

Reader, that evening was many, many years ago. I was then (almost) a young man. Heavy-handed Time has since obliterated from my features every vestige of youth, but has failed to efface one line which the next few hours wrote upon my memory's tablet.

I left Allen's Fresh about the dusk of evening. It had been cloudy and misty all day, and as night came darkly on, the clouds seemed to grow denser and the dampness more intense.

A gray fog, rising from the marsh below the village and floating up the swamp, wrapped in shrouds the trees whose motionless forms were growing dim in the gathering gloom.

It is strange that when we have passed through moments of such anxiety and mental strain, all things external to the one absorbing subject seemingly make not the slightest impression upon the mind, yet, even years afterward, when thought travels back to the past, we find that every incident and circumstance, no matter how trivial, connected with those moments, has left its impress upon the memory.

As soon as I was well out of the village I put whip to my horse and rode rapidly toward the spot where the man who was that night to test his fate was lying.

It was dark by the time I reached the place. I had never before visited the fugitives at night; I therefore approached with more than usual caution and gave the signal. Herold answered me and led the way to Booth. I informed them of what had just occurred at Allen's Fresh.

"The coast seems to be clear," I said, "and the darkness favors us. Let us make the attempt."

I then told them the safest way to proceed was for Booth to ride my horse and Herold to walk beside him, while I would precede them by fifty or sixty yards. When I came to a convenient place, I would pause and listen, and if the way seemed clear, would whistle. As soon as I gave the signal, but not before, they were to come forward till they reached the place where I was waiting, then stop there till I went forward again and gave the signal for their advance, and thus we would proceed to the river. If they did not hear my signal for their advance within a reasonable time after I went forward, I told them to get as noiselessly and as speedily as possible out of the road, and wait till they heard from me.

With difficulty Booth was raised by Herold and myself and placed upon my horse. Every movement, in spite of his stoicism, wrung a groan of anguish from his lips. His arms were then given to him, the blankets rolled up and tied behind him on the horse, and we began the perilous journey.

The route we had to take was down the cart track I have before mentioned, to the public road, a distance of about one mile and a half, then down the public road for another mile to the corner of my farm; and then through my place to the river, about one mile further, making the whole distance to be traveled about three and a half miles.

The part of our journey which lay over the public road was much the most dreaded by me; for not only were we more liable to meet some one on that part of the journey; but we had also to pass two dwelling houses situated close to the road; one occupied by a negro, named Sam Thomas, where there were children nearly always stirring round; the other was the home of Mr. John Ware, where there were several dogs.

The night had grown inky dark. No rain was falling, but the dampness clung to every thing and fell in drops upon us as we made our way among the trees.

As we journeyed cautiously on my feelings were wrought up to an intense degree of anxiety, not so much on my own account as for the successful accomplishment of what I had undertaken. When I paused to listen, the croaking of a frog, the distant barking of a dog, the whir of the wing of some nightbird as it passed over my head, would cause my heart to beat quicker, and my breath to come faster. When I gave the low whistle agreed upon as the signal that the road was clear, it sounded in my ears as loud as the blast of a trumpet, and though the ground was soft and yielding, the tramping of the slowly advancing horse, to my over-wrought fancy, was like the approaching of a troop. At length we reached the public road and entered upon the most dangerous part of our dangerous journey.

I walked softly down the road listening intently, but could hear no sound that indicated danger. I paused and gave the signal, and waited breathlessly as Booth and Herold entered the highway and came toward me. When they reached me I led the horse a few yards out of the road and told them to wait there. I then went on past Sam Thomas'. There was a light burning in the house that showed dimly through the mist, but I heard no one stirring. I was afraid to give the signal too close to the house, so I went on a little further than usual before I whistled. The house was safely passed.

Again the horseman and his companion waited and I went forward to the next most dangerous place, Ware's house. I walked past the gate and listened. Not a sound was heard. I moved a few yards further down the road and again gave the signal. As they came on by the house I expected every minute to hear the dogs bark, but they kept quiet. So far all had gone well. We were now nearing my place and would soon be off the public road and I began to breathe more freely.

At last, after what seemed an interminable age, we reached my place. We stopped under a pear tree near the stable, about forty or fifty yards from my house. It was then between nine and ten o'clock. "Wait here," I said, "while I go in and get you some supper, which you can eat here while I get something for myself."

"Oh," said Booth, "can't I go in and get some of your hot coffee?"

It cut me to the heart when this poor creature, whose head had not been under a roof, who had not tasted warm food, felt the glow of a fire, or seen a cheerful light for nearly a week, there in the dark, wet night at my threshold, made this piteous request to be allowed to enter a human habitation. I felt a great wave of pity for him, and a lump rose in my throat as I answered, "My friend, it wouldn't do. Indeed it would not be safe. There are servants in the house who would be sure to see you and then we would all be lost. Remember, this is your last chance to get away."

To refuse that appeal, prompted by a feeling I could so well understand, was the hardest thing I have ever had to do.

I entered the house through the kitchen. Henry Woodland was there. He had got in late and was just eating his supper. I asked him how many shad he had caught that evening and he told me. I then said, "Did you bring the boat to Dent's Meadow, and leave it there, Henry?"

"Yes, master."

"We had better get out another net to-morrow," I replied. "The fish are running well."

Some members of my family were in the dining-room when I entered. My supper was on the table waiting for me. I selected what I thought was enough for the two men and carried it out to them. None of the family seemed to notice what I was doing. They knew better than to question me about anything in those days.

After supper we resumed our journey across the open field toward the longed-for river. Though there was now little danger of meeting any one, I walked ahead, taking the same precautions I had used during the more hazardous part of our journey. Presently we came to a fence that ran across the path, about three hundred yards from the river. It was difficult to take it down; so we left the horse there and Herold and myself assisted Booth to dismount and supporting him between us, took our way carefully down the tortuous path that led to the shore.

The path was steep and narrow and for three men to walk down it abreast, one of them be-

108   J. WILKES BOOTH

ing a cripple, to whom every step was torture, was not the least difficult part of that night's work.

But the Potomac, that longed-for goal, at last was near.

It was nearly calm now, but the wind had been blowing during the day and there was a swell upon the river, and as we approached, we could hear its sullen roar. It was a mournful sound coming through the darkness. The most cheerful imagination could not have interpreted it into a voice of welcome and hope.

What it meant to the ears of the criminal trying to escape from the consequences of his crime, and about to trust himself to the mercies of the dark water — who can tell? If he heeded it at all, he might have construed it into an accusing voice upbraiding him for the blood he spilt at that very hour, one week before, upon its banks.

As we approached nearer the end of my part of the journey a thought suddenly occurred to

J. WILKES BOOTH   109

me with a shock; what if the boat was gone? I rapidly concluded if such a misfortune had befallen us what I would do. I would hide Booth and Herold in the almost impenetrable laurel growth that clothed the cliff we were then descending, and endeavor the next night to get the other boat that was hidden in the marsh around to Dent's Meadow.

At length we reached the shore and found the boat where Henry had been directed by me to leave it. It was a flat bottomed boat about twelve feet long, of a dark lead color. I had bought it in Baltimore the year before for eighteen dollars.

We placed Booth in the stern with an oar to steer; Herold took the bow-seat to row. Then lighting a candle which I had brought for the purpose — I had no lantern — and carefully shading it with an oilcloth coat belonging to one of the men, I pointed out on the compass Booth had with him the course to steer. "Keep to that," I said, "and it will bring you into Machodoc Creek. Mrs. Quesenberry lives near the mouth of this creek. If you tell her you come from me I think she will take care of you."

I then cautioned them to keep the light hidden and said "Good-bye."

As I was in the act of shoving the boat off Booth exclaimed, "Wait a minute, old fellow." He then offered me some money. I took eighteen dollars, the price of the boat I knew I would never see again. He wanted me to take more, but I said no, what I had done was not for money. In a voice choked with emotion he said, "God bless you, my dear friend, for all you have done for me. Good-bye, old fellow."

I pushed the boat off and it glided out of sight into the darkness.

I stood on the shore and listened till the sound of the oars died away in the distance and then climbed the hill and took my way home.

Though I knew my danger was by no means over, I felt that a tremendous load had been lifted from my shoulders. I had successfully accomplished what I had undertaken to do, and my sleep that night was more quiet and peaceful than it had been for some time.

On the whole I had been singularly fortunate during my attendance on Booth. I paid seven visits to him during the six days he was in the pines, but never met a single person either on my way there or returning home.

It is well known that Booth did not succeed in crossing the river that night. The strong flood-tide, against which I had forgotten to caution him, swept the boat up the river, and sometime during the night he and Herold landed at a place near Naugemoy stores, still in Maryland. They staid hidden somewhere in the neighborhood during Saturday, and at night succeeded in crossing to Virginia, and reached Mrs. Quesenberry's Sunday morning. Here they were met by my brother-in-law

Thomas H. Harbin, and a man named Joseph Badden of Prince George's County, Maryland, who did all they could to assist them, showed them a hiding place, carried them food from Mrs. Quesenberry's, and finally put them in charge of an old man of King George's County, named Boyan, who took them next day, which was Monday, on to Dr. Richard Stuart's.

The fugitives were not received at Dr. Stuart's, but were sent to a barn or out-house on the place. This treatment from Dr. Stuart was so unexpected by Booth that when food from the doctor's house was sent to him, he wrote a letter to the doctor, inclosing three dollars and saying that he would not accept hospitality when extended to him in such a manner without paying for it.

They left Dr. Stuart's place Monday evening in a vehicle driven by a negro named Lucas, and reached Port Conway on the Rappahannock River, where they crossed over into Carolina County the same evening.

Captain Jett and Lieutenants Ruggles and Bainbridge crossed the ferry from Port Conway to Port Royal at the same time Booth crossed and went with him to the house of Mr. Richard H. Garrett, where Booth spent Monday night. Tuesday, hearing Federal troops were in the neighborhood, he deemed it safer to leave the house and pass the night in a barn on the place.

About one o'clock that night a squad of Federal soldiers, commanded by Lieutenant Dougherty, came to Mr. Garrett's and demanded the man who had been staying there. They were informed by young Garrett that the men they were seeking were in a barn on the place. The troops surrounded the building and sent young Garrett in to Booth with a demand for his surrender. Booth refused to surrender and repeated what he had said often before, that he would never be taken alive. He offered to come out if the troops were withdrawn for a short distance so as to give him chance.

Lieutenant Dougherty then ordered the barn to be fired. This order brought Herold to terms. He surrendered and came out. The barn was then set on fire. Booth was seen in the light of the conflagration and shot through the head by Boston Corbett. He was taken to Garrett's house and laid upon the porch, where he died. His last words were: "Tell my mother I died for my country, and what I thought was best for it."

It was rumored at the time of Booth's death that his body weighted with a stone slab had been thrown overboard into the Potomac. I remember seeing a picture in an illustrated paper at the time, representing two men in the act of throwing the body from the boat into the water.

As a matter of fact Booth was buried in the inclosure of the old penitentiary in Washington. About four years afterward, when the prison was abolished and the walls pulled down, his remains were given to his friends and by them interred in that beautiful cemetery, Greenmount, in Baltimore City.

The boat in which Booth crossed the Potomac he gave to Mrs. Quesenberry and it was taken to Washington at the time she was arrested.

CHAPTER VII.

MY ARREST AND IMPRISONMENT.

The morning after I had taken Booth to the river Cox came to my house to say that, as there seemed to be no soldiers in the neighborhood, he thought it a good chance to get the fugitives across to Virginia. His surprise almost equaled his delight when I told him they had gone. Had I waited until Saturday night it would have been one day too late, as before evening the neighborhood was again filled with soldiers and detectives.

Sometime prior to the assassination of Lincoln a man named Carson, who was in Maryland on some secret business for the Confederacy, came to me to get put over the river. It was impossible at the time for me to put him across, so I hid him in Ware's pines, a short distance back of my place, until such time as it would be safe to move.

He had to remain in hiding several days and, thinking more of his own comfort than my safety, he built a kind of swinging booth of grape vines and boughs to lie in. When he went away he left behind him at this place a pair of socks and an old newspaper.

Some negro boys living on Ware's land happened to pass this spot while Carson was there and saw him. Of course they told what they had seen. And when Detective Franklin was searching for Booth, it got to his ears that Mr. Jones had been hiding some one in the pines. Franklin visited the spot and found the things Carson had left there. He sent two soldiers across to my house after me. These soldiers told me to saddle my horse and come with them. I inquired what they wanted and was told I would be informed of that by Detective Franklin, who was waiting to see me.

"Where do you want me to go?" I asked.

"Just behind the stable there, in the pines," was the answer.

"Then," I replied, "it is unnecessary for me to get my horse. It is a short distance and I can walk."

They insisted for a while that I should get my horse, but as I persistently refused they finally gave up the point and allowed me to walk.

When I reached the spot where Carson had staid, Detective Franklin pointed to the booth and asked me what it was.

"I don't know," I answered, "what you would call it."

"Well, what does it look like?" he inquired.

"It looks like the kind of place the farmers around here fix up to protect their hogs from the weather," I replied—as indeed it did.

"And I suppose," retorted the irate detective, "'the farmers' hogs wear socks, and read newspapers."

"Indeed!" said I.

"Look here," said the detective, sternly, "who built this place, and what was it built for?"

I was somewhat exasperated by my interrogator's manner and replied:

"It is none of my business who built it or what it was built for. This is not my land and I have nothing to do with it."

I was sorry next moment for my imprudence in giving way to temper.

"Seize that man and search him," was the detective's order, whereupon two of the soldiers caught hold of me and started to throw me down and pull off my boots.

"I will save you the trouble," I said quietly and seated myself. They searched me but found nothing. I was then placed under arrest and taken to Bryantown.

Realizing what would be my inevitable fate should my connection with Booth's escape become known, I did not feel altogether free

Bryantown Hotel

from anxiety during my imprisonment. And yet, I was not *very* uneasy, because from the questions that were asked me I was sure nothing was *known* against me. Besides, the very evidence that pointed to the fact that some one had been hiding in Ware's pines was against the presumption that it was Booth, because the socks were such as were worn by Federal soldiers, and the date of the newspaper showed it had been published *before* the assassination. About this time suspicion also attached to Cox. The negro Oswald Swann, who had guided Booth and Herold to his house, gave information against him, and swore, not only that the man with a broken leg went to the house, but that he also entered it. The latter charge was denied by Cox; and a colored girl named Mary, one of his servants, confronted Swann and swore that what he said was false.

I do not know whether Booth did enter Cox's house or not; I never thought to ask

Cox about it after all the trouble was over. If he did not, there must have been a light carried out for Cox to have been able to see the initials J. W. B., which he told me he had seen on Booth's wrist. But be the fact what it may, there is no doubt but that Mary's positive and persistent declaration that Booth had not entered the house—unshaken by threats or offered bribes—saved Cox's life when it hung by a thread.

Cox and I were kept prisoners in the old brick tavern at Bryantown; the same in which Booth had stopped when in southern Maryland eighteen months before.

While there we were subjected to every method that ingenuity could devise to make us divulge something. Sometimes some of the soldiers and detectives would get together under my window and describe how I would be hanged. I would look mournful but say nothing.

Cox and I slept in a room together, lying on

Henry Woodland

the floor, with our heads resting on our saddles. The first night we passed there—no one but ourselves being in the room except the two guards stationed at the door—as soon as we lay down Cox placed his lips close to my ear and whispered:

"What shall I do, Tom?"

"Stick to what you have said and admit nothing else," I answered.

The officer in charge at Bryantown was Colonel Wells and a rougher man I have never had to deal with.

About the second day after I had been taken to Bryantown, Cox was carried away, heavily ironed, and placed in the old Carroll prison, Washington. About one week afterward I was taken to the same place.

When I was taken from Bryantown I had to leave there my mare which I had ridden from home. I was anxious that she should be sent back, so when we stopped at T. B. on our way, to get something to eat, and to rest, I asked for a sheet of paper, saying I wished to write to a friend, Dr. Dent, and ask him to send my mare back to Huckleberry. Detective Franklin, who had me in charge, expressed his entire willingness that I should write to anyone I pleased, and put himself to considerable trouble to get a sheet of paper—paper happening to be a scarce article in T. B. that day. Of course I understand that this desire to accommodate me did not spring from any kindly feeling toward myself, but was intended to encourage me to write, in the hope that I might attempt to convey some secret intelligence to Dr. Dent, who was well known to be a warm Southern sympathizer and a fearless man.

I may as well add here that this mare was the same one Booth had ridden from the pines to the river that memorable Friday night. She was a flea-bitten gray, named Kit. Had her complicity in the assassination been known, what an object of interest she would have been.

Detective Franklin had some whiskey with him, which he freely offered to me. I did not seem to suspect his intention and would appear to drink. When he found he had failed to get me drunk he began to curse me and kept it up for the rest of the way.

I was at first placed in solitary confinement in the old Carroll prison. But through the clemency of Col. Wm. P. Wood this lasted but a few days.

Colonel Wood was a kind-hearted gentleman. He allowed me as many privileges and as much liberty during my imprisonment as was consistent with his duty. I remember with heart-felt gratitude his kindness to me during that period when friends who could assist me were few indeed.

John T. Ford, the manager of the theater in which Mr. Lincoln was killed; Junius Brutus Booth, brother to John Wilkes, and his brother-in-law, John S. Clark, the comedian; and Dr. Richard Stuart, were all in prison with me, charged with complicity in the assassination. I met these gentlemen daily.

Governors Henry A. Wise, of Virginia, Vance, of North Carolina, and Brown, of Georgia, and the Hon. Barnes Compton, of Maryland, were also my fellow-prisoners.

There was a certain time of day when the prisoners were allowed to walk in an inclosure adjoining the prison. I found out when Cox was in the habit of visiting this inclosure, and as I was not kept a close prisoner, contrived to meet him there.

He was not faring so well as I, but was still kept in solitary confinement. I found him low-spirited and despondent. Oswald Swann, the witness against him, was still in Washington.

After some weeks, though, Cox was treated with less rigor and we could meet sometimes without much difficulty.

One day, after we had been there about four weeks, Cox and myself were sitting together,

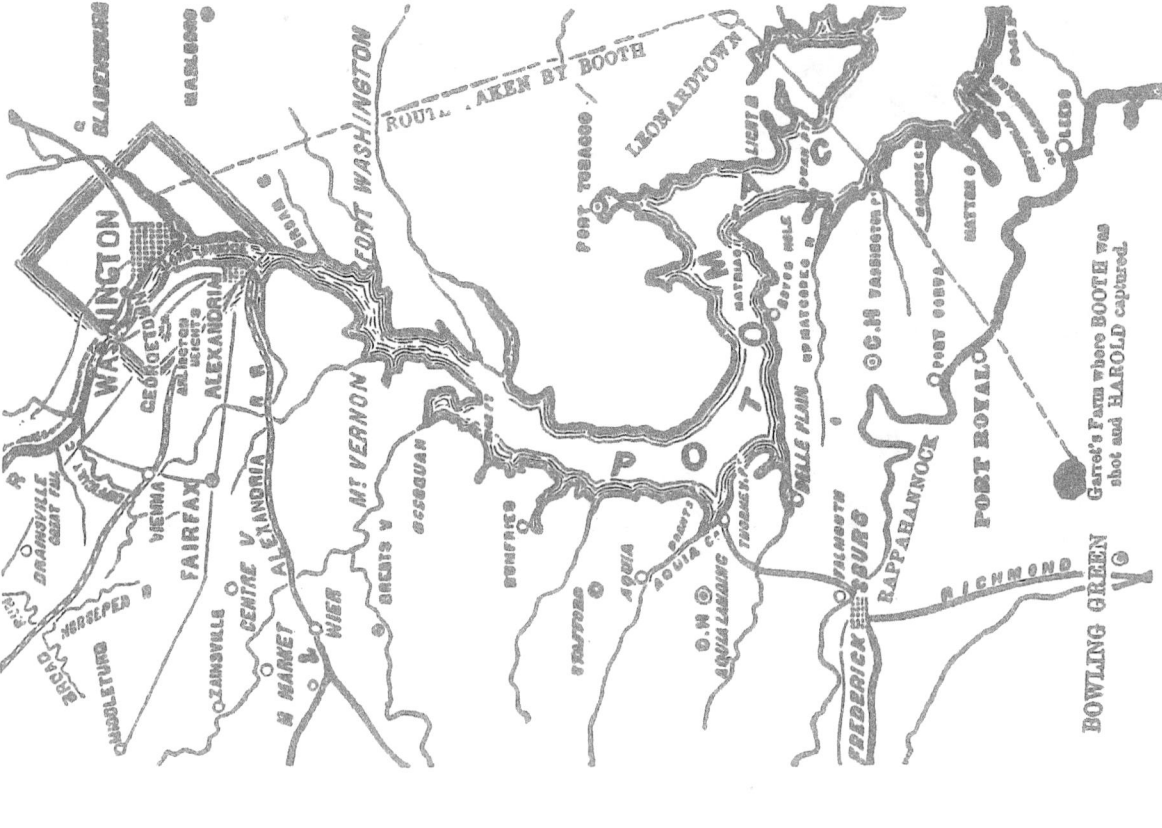

DEATH SCENE OF THE ASSASSIN.
Map showing where Booth was killed and Harold captured.

126   J. WILKES BOOTH

when I happened to look out the window and saw Swann, sachel in hand, going toward the Navy Yard bridge.

"You have nothing more to fear," I said to Cox, "the only witness against you has been dismissed and is going home."

About two weeks after this incident I was released; and one week later Cox was also liberated.

Reader, I have now told you the little part I played in the great tragic drama in which the men of a nation were the actors, and the fair fields of the South the stage.

THE END

PART 6

RETURN TO CHARLES COUNTY

In 1887 Jones returned to Charles County, apparently alone and still entrapped in penury. [54]

Back in Charles County Jones settled near La Plata, about a mile east of town. He purchased five acres near a spot called "Kitchen Gate," adjacent to land of Dr. Robert F. Chapman, on the public road leading from La Plata to Bryantown. The property was purchased with financial help of Jones's old friends, primarily Judge Stone. Collateral was in the form of two policies on Jones's life. La Plata attorney Adrian Posey presided at property settlement with Jones and Stone being the principals. Details of the mortgage requirements in the case lead one to feel Jones's long-time County friends were a bit less than generous with the old Confederate hero of Civil War times. [55]

Long political affiliations and old friend loyalties did secure for him the position of Justice of the Peace for the First Election District. This was a highly desirable position and a way to earn cash income when such was rare and elusive. About mid-June 1893 *The Crescent*, a relatively new La Plata weekly, announced that Thomas A. Jones and J. B. Mattingly had just come home from Chicago where they had contacted a printer and set a scheme in motion that would give them a publication about the Jones-Booth tale that would be sold at the Columbian Exposition. And Mattingly now was getting ready to travel southward to arrange for selling the new book in the old Confederate States. All looked promising at this point. Before the end of the summer the bottom fell out completely and Jones once again faced grim disappointment when the newly printed books had to be brought home for storage in a La Plata barn.

In 1901, a fascinating book titled *Assassination of Abraham Lincoln* came off the press in Washington, D. C. It was authored by Osborne H. Oldroyd who was then one of the most active, best known Lincoln assassination story expert and a collector of Ford's Theater-John Wilkes Booth memorabilia. Shortly after the Civil War Oldroyd, who had been a Union Army volunteer infantry officer with an Ohio regiment, settled in Washington, D. C. searching out stories and artifacts about the Ford's Theater final act in a President's life. Sometime before 1900, he purchased the Peterson house where Abraham Lincoln had been carried to his fading earthly hours. In this house Oldroyd kept his Lincoln memorabilia open to public view. In preparing material for his 1901 book he, now over 60 years of age, walked the entire Booth-Herold escape trail from the alley behind Ford's Theater to the Garrett farm near Port Royal, Virginia. In the walk through Charles County he met a few survivors of that April 1865 fugitive visitation

---

[54] *Port Tobacco Times*, September 23, 1887, Volume XLIV, Number 15 included the name of Thomas A. Jones as a newly registered voter in the County's First Election District... list of September 6, 7, 8, 9 and 10, 1887. This district covered the heart of the county including La Plata and Port Tobacco.

[55] Charles County Land Records, JST 1, folio 354.

including Henry Woodland who had helped Thomas A. Jones get the assassin and his companion launched into the treacherous Potomac River currents.

Oldroyd described a meeting with Thomas A. Jones earlier at the Peterson house. He wrote,

> *"In the month of April 1894 an old man, 74, called at the house where the martyred president died. After viewing the room he said: 'My name is Thomas A. Jones and I am the man who cared for and fed Booth and Herold while they were in hiding, after committing the awful deed.' He was asked to tell the story and Oldroyd relates what he remembered about what Jones told him, all quite in accord with the Jones 1893 Chicago publication. After Jones had described having heard from Detective William Williams about the government's offer to pay $100,000 for Booth, dead or alive, Oldroyd interrupted and asked Jones if it was any temptation, being in the reduced circumstances he was at the time. He proudly answered: 'No, indeed; my word could not be bought for a hundred times that amount. I considered it a sacred trust. The little I had accumulated was irrevocably lost, but, thank God, I still possessed something I could call my own, and its name was honor.'"* [56]

Samuel Cox's servant Mary Swann Kelly probably saved the lives of Cox and Jones when she contradicted the story of Oswald Swann about the kind of reception Cox gave Booth and Herold when they reached "Rich Hill." That Cox did not admit them into his house as friends was crucial to the Federal case against him and Jones. Both went free after Mary's testimony.

---

[56] Osborne Oldroyd, *The Assassination of Abraham Lincoln*, Washington, D. C.: 1901.

Jones continued to be Justice of the Peace for the county's First Election District until his death. According to a *Crescent* news note on May 18, 1894, Jones had recently committed to jail a couple of tramps lounging around the district and the sheriff had taken them to the House of Corrections. A few days later the *Crescent* reported that Jones was driving home from La Plata in his road cart, when a piece of lumber on the seat beside him slipped down on his horse's heels. This caused the frightened horse to run and Jones was thrown off the cart and into a roadside ditch, sustaining only trifling scratches. Less than a year later the *Crescent* informed its readers in the March 1, 1895 issue that.

> *". . . Justice of the Peace Thomas A. Jones of John Wilkes Booth fame is lying dangerously ill of heart trouble at his home near this place. He is 75 years of age, and his death is expected at any moment."*

Of the Jones obituaries in the three Charles County weeklies the one in the *Crescent* of March 8, 1895, gave the best account of Jones's long, troubled life. He was

> *". . .guided by the highest sense of honor . . . his zeal, fidelity and courage amply proven . . . won for him the praise of all who esteem fidelity and honor between man and man . . . there is no one who does not believe that Thomas A. Jones acted the part of a hero."*

Such were the accolades accorded Jones by his distinguished neighbors when he died.

"NECROLOGY
Thomas A. Jones

> *Mr. Thomas A. Jones, of this county, died at his residence near La Plata Saturday night last, after an illness of more than a week, from congestion of the lung aggravated by extreme old age and complication of other troubles.*
>
> *Mr. Jones was widely known, not only throughout this county and State, but throughout the country on account of the prominent position he held during the Civil War as a secret service agent of the Confederate Government, and his connection with John Wilkes Booth and David C. Harold [sic], during there [sic] sojourn in this county after the Lincoln assassination.*
>
> *Thomas A. Jones held important positions in many daring adventures during the war period, and his zeal, fidelity and courage have been amply proven. He was the soul of honor. His connection with the escape of Booth and Harold, while it may be*

viewed differently by the partisan of either side with regard to propriety, has won for him the praise of every one who esteems fidelity and honor between man and man. It has won for him a prominent place in the history of this county; and whatever may have been the effect of his act in shielding a criminal, there is no one who does not believe that Thomas A. Jones was guided by the highest sense of honor and that he acted the part of a hero.

The deceased was born in Port Tobacco October 2, 1820.

At the time the civil war broke out he had accumulated enough to enable him to live comfortably, but his property, consisting largely of slaves, was lost when peace was restored. For a number of years after the war he held office either under the state or general government. He was employed in the Washington navy-yard, and was displaced during Harrison's administration. After that he was made Justice of the Peace for the First District of this county, which he held at the time of his death.

He kept his secret of helping Booth and Harold to escape from Southern Maryland into Virginia for nearly twenty years. It was well he could keep a secret and that no one betrayed him. For he was arrested and imprisoned in the old capitol at Washington after the assassination, but the Federal authorities were unable to connect him with the affair and had to discharge him. He never spoke of the adventure even after he broke his long silence unless the subject was introduced by some one else. He first told the story in connected forms for publication in 1884. In June, 1893 he published the whole narrative of the sojourn of Booth in Southern Maryland after the assassination in a book, which John E. Stone, Esq., a member of the Bar of this county, aided him in writing.

Mr. Jones was a brother of Mrs. Mary Hindle of this town. He was married twice. By his first wife, a Miss Harben [sic], he leaves nine children, Richard, John, William, Henry Clay, Jane F., Alice, Bettie, Annie and Mrs. Charles Penn.

By the last wife Master Edward Jones, a youth, survives him. The interment, under the management of Mr. C. E. Wade, took place at Newport Catholic Church on Tuesday, in the presence of a large assemblage of friends. The active pall-bearers were J. H. Roberts, R. F. Mattingly, J. Samuel Turner, W. H.

Wenk, J.E. Mudd and L.A. Wilmer. Rev. J.E. Wade officiated." [57]

The late prominent southern Maryland historian Robert E. T. Pogue admired Thomas A. Jones greatly. In his fine work titled *Old Maryland Landmarks* the author shows he did more than a little research about the Lincoln assassination story as it related to southern Maryland places and people. He relied heavily on GATH's 1884 account of the Booth-Herold-Jones experiences. In summing up the nature and character of Jones, Pogue asks,

> *". . . And now, fellow tourists, what is your opinion of Mr. Thomas A. Jones? Have you ever known anyone like him? What manner of man was this who risked his own life and property to take care of a murderer with whom he had never been acquainted, and owed no favors to? How many would turn down an offer of $100,000 for a wanted criminal? I guess the only explanation is Jones's own . . . he just felt sorry for him."*

The enormity of Jones's sacrifice really must be measured in terms of today's comparable dollar purchasing power. In 1865 any successful southern Maryland farmer would have thought he'd had a good year if he ended up with $200 free and clear at the end of the harvesting and marketing of crops produced in an average year on about 250 acres. Samuel Cox, Sr., with about $30,000 worth of cash, real estate and farm equipment, would have been considered a very wealthy man . . . GATH doubtless figured this correctly.

Pogue ended his evaluation of Jones the man with these comments:

> *"In my opinion Jones was a shrewd, intelligent, fearless and loyal little man. Yes, loyal to the South, loyal to his friends, and loyal to this miserable, injured, misguided man who had committed a terrible crime. Once he had made up his mind, he was capable of nothing but loyalty. I would like to have such a friend as Thomas A. Jones."*

And what greater commendation can anyone make of another human being? The Jones risks and sacrifices were matched only by those of his immediate large family, who in 1865 were all under twenty years of age. And all had suffered under an ever-diminishing standard of living and much reduced security in every way following the outbreak of the war. [58]

---

[57] Within half a century the final resting place of Jones at St. Mary's Church at Newport could not be located. Grandson Raymond A. Jones said about 1988 that he and his father Ernest E. had marked the grave site with a "cedar stob" about 1920 . . . a rather unsubstantial memorial for a man whose life works left tracks in American history with reflections that endure even now coast to coast for those fascinated by the overall scope of the Lincoln assassination tragedy . . . fact or fiction.

[58] Robert E. T. Pogue, *Old Maryland Landmarks*, (Bushwood, Maryland: 1972).

Than Whom a Truer Man Never Stepped the Earth . . . this magnanimous accolade for Jones came from his wartime commanding officer, "late Colonel & Chief of Scout Service, Confederate States of America," William Norris of Maryland. In a letter of September 25, 1890, Norris told Mr. Eugene Diderly of Baltimore (apparently a newspaper reporter] how to get in touch with Jones who then lived in southeast Washington, D.C. Jones was described as

> *". . . a most estimable man - gentle, modest, retiring and absolutely reliable . . . a gentleman in every fibre . . . one who knows a great deal about the [CSA] wartime underground in Maryland."*

Norris knew of Jones's plan to write a book about his 1865 experiences with Booth and Herold. He indicated that if it was not a success financially he would help Jones get into the Confederate Veterans' home in Pikesville, near Baltimore. [59]

When the Norris - Diderly letter was written Jones had just been dismissed from what probably was a patronage job at the Washington Navy Yard . . . one he doubtless had held for only a year or so. When Jones moved back to Charles County about 1887 he became a resident of the Fifth Congressional District served by an old acquaintance, Honorable Barnes Compton, who had been elected in 1884 as a Democrat and served until March 20, 1890 when he was succeeded by Charles County Republican Sydney E. Mudd in mid-term because of election irregularities in 1888. As patronage matters went in those days the incoming Republican would have lost no time in replacing Compton friends with supporters of the new member of Congress. And Jones, a life-long Democrat in Charles County, could not have been surprised to find himself unemployed after the Mudd victory.

The following verse was published in a Charles County weekly newspaper shortly after Jones died and it quite likely was a tribute to him:

THE MEN FOR THE TIMES
  Author Unknown

> *Give us the nerve of steel,*
> *And the arm of fearless might,*
> *And the strength of will that is ready still,*
> *To battle for the right.*
>
> *Give us the eye to weep;*
> *That honest tear of feeling*
> *That shuts not down for the world's dread frown*

---

[59] The Chicago Historical Society, Chicago, Illinois letter from William Norris to Eugene L. Diderly, Sept. 25, 1890. Carried in the *Surratt Courier*, Vol. XII, No. 11, November 1987: Clinton, Maryland.

The genuine heart revealing.

Give us a mind to feel
The sufferings of another,
And fearless power in the dying hour
To help a suffering brother.

Give us the clear, cool brain
That is never asleep or dosing;
But springing ever, with bold endeavor,
Wake the world from its prosing.

Ah, give us the nerve of steel,
And one hand of fearless might,
And the heart that can love and feel,
And the head that is always right.

For thos foeman is now abroad,
And the earth is filled with crimes;
Let it be our prayer to God,
Oh, give us the men for the times.

The 1895 October Term of the Charles County Court disclosed a sad picture of the worldly worth of Jones when he died. Attorney L. Allison Wilmer was administrator of the estate, and he reported an inventory dated September 3, 1895. The entire estate amounted to $271.70, including funds from a $5.45 sale of poultry and a paid-up life insurance policy with New York Life. Claims against the estate brought the balance down to $181.60. [60]

But we cannot measure only in dollars here. Certainly the overall, great balance sheet of life must show for Thomas A. Jones an incalculable profit that sometimes accrues to the memory of great and courageous men no matter what their political persuasions or philosophies of life.

---

[60] *Archives of Maryland*, Annapolis, Maryland, Orphans' Court of Charles County, Maryland, 1895 October Term, signed by J. Benjamin Mattingly, Register of Wills.

Jones stated at least once that he had no stauncher, more loyal friend than Henry Woodland. This Black man lived on the Jones farm for years and played a major role in helping protect the assassination fugitives and getting the boat ready for the dash across the Potomac. This never before published photo of Woodland was taken at his Pope's Creek home in 1922 when he must have been about eighty years old. Photo courtesy of Cynthia Q. Wilmer.

PART 7

SMOOT'S GREAT STURDY BOAT

A provocative, tantalizing sidelight on the Booth-Herold-Jones escape adventure was told in a story written about 1900 by Richard Mitchell Smoot from Port Tobacco. It was titled quaintly and illogically, *The Unwritten History of the Assassination of Abraham Lincoln*. Printed in Baltimore early in 1904, the publication died a-borning. It was produced by the John Murphy Company but just before copies could be bound and shipped, the disastrous Baltimore fire of February 7, 1904 destroyed the printing plant and all copies of the first edition. Fortunately, the first five copies were bound and sent to the author for copyright registration in Washington, D. C. Three copies were presented to friends of Mr. Smoot.

The author died May 8, 1906 in Little Rock, Arkansas without making new arrangements for publication of his work. Shortly thereafter, the author's daughter sent the manuscript to Orra L. Stone, a journalist in Clinton, Massachusetts. Stone printed the manuscript late in 1908. It was a very brief work of about 25 pages, now long out of print. Richard M. Smoot had farmed in Charles County near Port Tobacco, and when the Civil War made farming unprofitable. He turned to conducting a ferry service across the Potomac River for man, beast and merchandise. He wrote that his boat was large and stout and used intensively in Potomac River blockade running until late 1864. Oddly, nowhere in his sparse little tome does Smoot even mention Thomas A. Jones. By today's distance reckoning, they probably lived about 10 miles apart--the distance from Port Tobacco to Pope's Creek.

Smoot apparently knew a great deal about two interesting aspects of the Lincoln assassination-kidnaping plot. John H. Surratt visited Smoot early in 1864 to try and buy Smoot's rather large, sturdy river craft. During this visit Surratt indicated he could use two other boats if each could hold at least 15 people with a capability of crossing the Potomac quickly and safely. Shortly thereafter, having agreed to accept $250 for the boat, Smoot and Surratt closed the deal in the Port Tobacco law office of attorney Frederick Stone where Smoot received $125 and Stone was given $125 to keep in trust for Smoot. Now, as instructed by Surratt, Smoot turned his boat over to George Atzerodt and he in turn placed it in charge of George Bateman, a farmer living near or on King's Creek which flowed into the Nanjemoy Creek on its east side. There it was hidden waiting for a time when it would be urgently needed by certain shadowy persons conspiring to commit some unknown dark and deadly deed.

Several months passed before Smoot heard anything more about the plotters nor received any more money. Then, things began to fall into place for him. After visiting Mary Surratt at her "H" Street home about 9 p.m., April 14, 1865, Smoot fled the city at the suggestion of Mrs. Surratt and spent the night in Alexandria, Virginia. There, very early the next morning, he found out about the Lincoln assassination and heard named those intimately involved in the crime. Doubtless, he felt at this time that at long last his boat would be needed and he might get the $125 still owed to him. At the Surratt house the evening before, he had been assured by Mary Surratt that the boat would be used that night.

And, now, to show the fickleness of fate and fortune, because of John Wilkes Booth's leg injury after the assassination, the Smoot boat was not used by the conspirators. In fact, George Bateman chopped it into tiny pieces and burned them upon hearing about the assassination. So, the story of the Thomas Austin Jones involvement with the fugitives in their remaining lifetimes, all hinged on John Wilkes Booth's desperate need for the help of a doctor late on April 14th and at this time the only sympathetic one he knew practicing medicine in southern Maryland was Dr. Samuel A. Mudd who lived near Bryantown. Booth had met this man and at least once had visited his home and so knew exactly where to go. Now the stage became set for the involvement of Mr. Jones of Pope's Creek in the Lincoln assassination-Booth escape tale.

In his brief account of 1904 Smoot wrote that Judge Stone in Port Tobacco gave him the $125 balance for the boat purchase between one and two months after the Lincoln assassination. Smoot stated that the *"enforced change of plans prevented the conspirators getting to where his boat was and they were so hard pressed that they were compelled to put up with the first boat they found at hand."* And that was the little slate-gray skiff used by Booth and Herold provided by Thomas A. Jones.

The second Lincoln-assassination-related matter that Smoot knew about intimately had to do with the 1867 trial of John H. Surratt, Jr. He had been extradited from Egypt and many thought he would get his due for having been deeply involved as an assassination conspiracy planner. As we know now, John Surratt got off lightly and according to Smoot the reason was that Smoot's brother Ned was offered a bribe by the government prosecuting attorney handling the case. Ned was to testify on behalf of the government about what he and perhaps his brother, Richard, knew about Surratt and the big sturdy boat affair. Ned Smoot, as soon as he could get away from the prosecuting attorney, went to Washington, D. C. attorney William Matthews Merrick, a Charles County native, who was defending Surratt and told him about the attempt to bribe. At the trial Ned said nothing to help the government's case and when turned over to Merrick for cross-examination, blew the whistle on the government prosecutor with respect to the bribe. It was not denied by the government prosecution. And, as Smoot wrote the *"effect of my brother's statement on that jury was almost palpable. Surratt and his friends saw in it the foundation of a strong hope for a disagreement, at least."* And of course, in fact, John Surratt was acquitted. [61]

The Smoot kidnap plot boat story is included here in part because of its oblique connection with the Jones' story, and in part because it deserves more attention by historians than it has been accorded. However, the Smoot account is supported by more substance than many

---

[61] Richard Mitchell Smoot, *The Unwritten History of the Assassination of Abraham Lincoln*, John Murphy Co., Baltimore, 1904 (reprinted by Orra L. Stone, Clinton, Massachusetts, 1908).

have previously supposed. Those who know something about the history and heritage of southern Maryland understand that Smoot's relation of events in which Judge Frederick Stone and Judge William Matthews Merrick were involved enjoy considerable validity. These names even today remain among the most respected of those connected in special and honorable ways to southern Maryland's stance during and after the "War Between the States."

Stone, who died in 1899, was one of the most distinguished jurists of Maryland for half a century, and served as a Congressman for two terms.

Judge Merrick's father was Senator William Duhurst Merrick, a Charles County hero of the Battle of Bladensburg and a Charles County Register of Wills about 1838-40. Son William Matthews Merrick had been a prominent jurist in Washington early in the war and had been unjustly treated by Federal military authorities for objecting openly and strenuously to suspension of the protection of writs of habeas corpus by the Military Government of the District of Columbia. In defending John H. Surratt, Judge Merrick once again faced Federal authority in court. Now he was teaching law at Columbian College and would end his teaching career as a member of the law faculty of Georgetown University. He had only increased his stature as a man of the law by standing up to Federal military martial law excesses during the war. [62]

President Cleveland returned Merrick to the bench of the District of Columbia Circuit Court in 1885, from which he had been driven by the wartime D. C. military authority. The judge served Maryland as a Democrat in the 42nd Congress. [63]

Richard Mitchell Smoot's inclusion of such notable personalities in his story in important roles must impart a great helping of credibility to his work. After all, Smoot himself lived in Charles County until after 1900 . . . and his neighbors continued to be the Stones, Matthews and Merricks, all with many years of unbroken political and social power and prestige in southern Maryland. Smoot himself could boast about long, most respectable lineage in Maryland, going back before the Revolution, including an officer serving in the Charles County Revolutionary War militia.

Smoot remained in Charles County many years after his traumatic evening in Washington April 14, 1865. A society note in the January, 18, 1884 issue of the *Port Tobacco Times* reported that George W. Cross of Little Rock, Arkansas had married Cordelia Smoot. [64] She was a daughter of R. M. Smoot, keeper of the Cedar Point lighthouse. At this time a son of R. M. had lived in Arkansas for a while and sister Cordelia may have met her future husband during a visit to her brother's home. Richard Smoot and his wife moved to Fort Smith, Arkansas shortly after

---

[62] *Georgetown College Journal*, Georgetown University, Washington, D.C., April 1885 and February 1889 issues.

[63] John M. Wearmouth, *Charles County Helps Shape the Nation*, (La Plata, Maryland, 1986), pp. 35-37.

[64] Roberta J. Wearmouth, *Abstracts, Vol. 4: 1876-1884*.

1900 to live with their son. Mary E. Smoot died there on January 2, 1904 and Richard M. Smoot died May 8, 1906. [65]

The person to whom Smoot originally related his story is unknown. And where told remains a mystery as well, although the printing in Baltimore indicates the Smoot account was heard first in Maryland, after the death of certain distinguished Charles County actors on the 1865 stage of the John Wilkes Booth escape adventures. [66]

Barnes-Compton house on main square in Port Tobacco about 1873 sketched by George Alfred Townsend. Small building on left housed carriage repair and painting shop operated by Atzerodt brothers - - George was a Lincoln assassination conspirator hanged in 1865 for his complicity.

---

[65] Ibid.

[66] Curious about who Orra L. Stone was and how Smoot's daughter met or heard of him, the authors spent a day in Clinton, Massachusetts (about 40 miles west of Boston) late in 1994. Clinton Historical Society people knew much about Stone but nothing of Smoot's Lincoln assassination item explained here. Stone died in 1961 at age 88 after a long lifetime in local publishing. He was 35 years old when he printed the Smoot work. For 35 years he worked for the *Clinton Daily Item*. Oddly, almost as an avocation O. L. Stone practiced law after graduating from Boston University's School of Law cum laude in 1901. It was while a lawyer that he used connections with a local printing shop to produce the R. M. Smoot sidelight on the Lincoln assassination affair. Surely Stone, an avid Civil War historian and son of a prominent local veteran of that conflict recognized something valid and credible in the writing about Smoot's connection with the 1864 plot to kidnap the president, with first stop toward Richmond being the sultry marsh-bounded King Creek in Charles County that flowed into the Nanjemoy, which in turn joined the Potomac about half a mile southward.

JOHN H. SURRATT.

Among the most interesting events of the last month was the arrival at Washington of the United States steamer *Swatara*, having on board JOHN H. SURRATT, the alleged accomplice of JOHN WILKES BOOTH in the assassination of President LINCOLN on April 14, 1865. After a flight to Europe and Africa, occupying nearly two years of time, the wretched criminal has been brought to the scene of the assassination for trial. The trial of the Conspirators at Washington in 1865 developed the fact that SURRATT was the principal accomplice and dependence of BOOTH in carrying out his infamous designs; and a large reward was offered in the hope of securing his arrest. SURRATT had, however, fled; he was probably in Canada when the murder was consummated. The first intimation which the Government received of his whereabouts was by a letter, dated September 27, 1865, written to Secretary SEWARD by A. WILDING, Vice-Consul at Liverpool. SURRATT, it seems, remained in Canada four months, having been secreted most of the time by a Roman Catholic priest at Three Rivers. He was disguised, having dyed his hair, eyebrows, and mustache, stained his face, and put on glasses. It appears that SURRATT had manifested no signs of penitence, but had told his friends that, if he could live two years longer, he would serve JOHNSON as he had LINCOLN.

He arrived in Liverpool on September 25, 1865. On the passage from Quebec to Liverpool he traveled under the name of M'CARTY, and was introduced to several persons as "a Confederate who had compromised himself." To one fellow-passenger he became quite confidential; spoke of having great difficulty in escaping from the United States to Canada; smiled when some connection with the assassination of LINCOLN was imputed to him; admitted that he had been in the rebel service, engaged in conveying intelligence between Washington and Richmond, and that he was concerned in a plan for carrying off LINCOLN from Washington, which was entirely concocted by BOOTH and himself, and that he came to Canada just before the assassination; and, finally, he de-

JOHN H. SURRATT.—[SKETCHED BY A. M'CALLUM.]

SURRATTSVILLE, THE HOME OF JOHN H. SURRATT.

PART 8

REFLECTIONS

In all the Jones-Booth saga perhaps the most trying time for Thomas A. was his final ordeal of imprisonment "on suspicion." No writer ever has covered this subject better than Jones did in his 1893 *J. Wilkes Booth*. Surely he was aware that his last view of Booth and Herold in the slight skiff disappearing into the Potomac nocturnal murk could not eliminate the threat of Federal apprehension. Washington could not help but remember Jones's Confederacy ties during 1861. And Jones probably continued to keep up with Government man-hunt operations through contact with military forces still sifting through most of southern Maryland, not knowing yet the route of Booth's escape trail.

Washington journals reflected the increasing frustration of the government over losing Booth and Herold. About April 20 Stanton ordered distributed proclamations offering rewards in amazingly generous amounts and carrying the words, *"The Murderer of our Beloved President is still at large . . ."* and appealed for *". . . the stain of innocent blood be removed from the land "* and exhorted *"all good citizens to aid public justice . . . "* Certainly this proclamation was tacked up on major public buildings throughout southern Maryland. And at their reading all those Charles Countians who had helped Booth and Herold in any way must have suffered a few qualms and regrets. [67]

The following words by Herman Melville create even today an impassioned feeling for the mood of many following Lincoln's death:

> . . . There is sobbing of the strong,
> And a pall upon the land;
> But the people in their weeping
> Bare the iron hand:
> Beware the People weeping
> when they bare the iron hand.

And, within just days after Jones dispatched his fugitive guests, a peculiar combination of circumstances led Federal searchers to "Huckleberry."

Cox, too, had been picked up because Oswald Swann told government authorities that Cox had entertained the fugitives several hours in his house April 16. Cox denied this and was supported by a servant, Mary Swann. Because of her testimony and lack of any other substantial reason for holding Jones and Cox the two were released after about six weeks. [68]

---

[67] Margaret Leech, *Reveille in Washington, 1860-1865*, Harper & Bros., (New York: 1941), pp. 406-407.

[68] Thomas A. Jones, *J. Wilkes Booth*, Laird & Lee Publishers, Chicago, 1893.

Incredibly, the assassination conspiracy investigation network never again intruded into the lives of either Cox or Jones. In weeks and months following, about ten others were either executed or sentenced to years in prison for less reason than there would have been to so penalize Cox and Thomas A. Jones. Doubtless both reflected on this fact for the rest of their lives. And, surely, Jones never dreamed in early post war years that someday he would write a book on the subject of his aid to Booth and try to sell it in Illinois . . . perhaps even to old troopers of the 8th Illinois Cavalry, which had disturbed much mud and generated great clouds of dust along southern Maryland roads in April 1865 on a fruitless, rumor-fed search for the arch conspirators.

Face to Face With Captain Williams . . . Again

A fascinating re-hash with new slants on the Jones-Booth story was the sparkling product of a visit Thomas A. had in Washington with an old and once formidable adversary about 1890..

Most people, even Marylanders, know little more about Jones than what appears boldly on a historical marker erected years ago in a now moribund Port Tobacco. Here, in the center of Jones-J. Wilkes Booth country, about 100 feet from the site of old Brawner House, are a few words in metal that tell much about the total fabric of Jones. These words have for years been etched into the minds and memories of many. They are a brief account of perhaps the most trying, threatening experience Jones had during his total war-years support of the Confederacy. In the dingy basement bar room of the Brawner House Federal Government detective Captain Williams and Thomas A. Jones had a chance encounter a few feet from each other on about the evening of April 17, 1865. Here, while each perhaps nursed a shot of locally produced "white lightning," the detective announced that the government would give at least $100,000 for information leading to the capture of Booth and Herold. Jones disclosed nothing . . .and he knew so much. (In 1970 the authors stood upon the ancient herringbone laid brick of this cellar bar room while doing archeological investigation of the hotel site.)

A quarter of a century later Jones and Williams talked about that moment in the little bar room in Port Tobacco. A Washington, D.C. newspaper man captured the scene in the detective's F Street office about mid-1890 while Jones was living on 12th Street in southeast Washington. He seems to have moved into the Nation's Capital after establishing his new home in La Plata. Jones had been given a patronage type appointment in the Washington Navy Yard, but this ended abruptly when Sydney E. Mudd, Senior, replaced Barnes Compton as the Representative of Maryland's Fifth District. This Jones-Williams meeting seems to have taken place shortly after Jones lost his government job.

Following are selected direct quotes from the article.

According to Williams, about daylight of April 15 he was ordered by Washington Provost Marshal O'Beirne to head directly for southern Maryland with a group of cavalry. Williams had asked O'Beirne, "Where must I go?" And O'Beirne had answered,

"How do I know?" followed by "Go and don't return to Washington until you find Booth, but mind, don't harm a hair of his head!" Williams and horsemen soon crossed Eastern Branch bridge, where Williams in a frantic rush knocked down a sentry by running over him with his horse.

The search expedition first stopped at Surrattsville to talk to John M. Lloyd who was running the tavern. It was here that Mrs. Surratt had left field glasses and carbines for Booth and Herold. Lloyd was sent back to the City under arrest. And, said Williams to Jones, ". . . from here we went to Bryantown," and, said the captain eyeing Jones closely, "Of course I remember you. I can never forget that come-to-the-Lord-and-be-saved expression you wear now and wore then. But if I had known then what I do now, how different would things have been! Why, you ought to be shot! If you had told me where Booth was you would have been the biggest man in America, and would have had money by the flour barrel full."

And Jones replied to the old detective . . . "Yes, and a conscience black as purgatory, and the everlasting hatred of the people I loved. No, captain, I never the first time thought of betraying Booth. After he was placed in my hands I determined to die before I would betray him." Then Williams asked, "Who placed him in your hands?" Then Jones, "Samuel Cox . . . " Jones then went over the story of what followed, including the launching of the gray skiff at the Potomac bank. Williams remarked to Jones that if he had disclosed Booth's whereabouts he would today be General Jones instead of a discharged laborer from the Navy Yard. Jones retorted, "That may be true, but how could I give up the life of that poor devil over there in the pine thicket hovering between life and death, and the confidence reposed in me by the best friend I ever had, Samuel Cox? I was a creature of circumstances. I did not know Booth, but when Cox put him in my keeping nothing would have tempted me to betray him." [One must wonder here . . . where did the greater loyalty lie . . . for Cox or Booth?] *Jones continued, "I have lived in plenty and I have lived in poverty, but God knows I have never betrayed a trust or done that which I believed dishonorable . . . but for the mean little spite-work of Congressman Mudd this matter would have never come out."* [Actually, it was far too late for silence . . . George Alfred Townsend had seen to that in 1883 and 1884.]

William Williams, Federal detective who headed search for Booth and Herold from Port Tobacco headquarters.

Jones asked Williams if he remembered the imprisonment of himself in the tavern in Bryantown. Williams did remember and said, "Yes . . . and let me say to you that myself and other officers believed that you knew more than you would tell, but that sanctimonious look of yours saved you." Jones told the detective that while locked up at Bryantown he overheard Union soldiers discuss the later stages of the Booth escape story and they seemed to have no inkling about what really was happening just before Booth's death. Jones said he had been cursed and abused at Bryantown before being sent to Washington in an ambulance in charge of a man who tried very hard to get him intoxicated during the trip . . . hoping of course to loosen the tongue a bit . . . very difficult to do with Thomas A. He did allow in the course of talking with Williams that he was very upset when he heard about the broadsides being posted everywhere that promised that anyone who even furnished bread or water to the assassin and friend might be put to death in consequence therefore. This threat apparently did shake this calm, quiet, sturdy man with his incredible store of inner strength and resourcefulness.

Jones swore to Williams that he knew nothing about Booth's plan to murder Lincoln. But he indicated he was long aware of the kidnap plot and said all was in readiness in Charles County . . . boats and men. He felt that only the desperate public road conditions made the kidnap plot untenable in late 1864. When Williams asked Jones, "You were in the secret service of the Confederacy?" he was told, "I was the Chief Signal Agent of the Confederacy north of the Potomac." Jones stated he took great chances of being killed doing this work. His relation of the trip to Richmond to collect his pay for years of extremely demanding service makes a very sad story. Richmond was soon to fall . . . all was confusion and uncertainty. The chief signal agent failed to collect the $2,500 owed him. He had unwisely allowed his pay to accumulate for most of the time since 1861. Even worse, he lost over $3,500 in CSA bonds. He said to the detective, "It all went and I was left penniless. The war was a very bad thing for me a ll the way through."

After Jones left the detective's office the reporter was told by Williams, "I have dealt with and sized up many men during my life, but that man Jones beats them all. He has changed very little during the past 25 years. This is the first time I have seen him since we met at Port Tobacco and Bryantown, and yet I remember every feature. He is a wonderful man, and one that, when he believes he is right, nothing can change him. I remember when I made that offer of $300,000 [sic] in the saloon he was standing next to me at the bar and I could not detect the least movement or change of his face. There was something which told me he knew where Booth was, or could give us information which could lead to his capture, but he couldn't be worked. No amount of money or glory would have tempted him. No human being can read his face and tell what is passing in his mind. It is like a stone. He would have gone the hemp route if the facts he now gives had been known then. If he had only told me where Booth was, Boston Corbett would never have had a chance to shoot Booth. We wanted him alive!

*This remarkably reported conversation ended with Williams saying, " . . . the assassins had crossed Eastern Branch bridge, at Lloyd's Tavern Herold had joined Booth and the two had proceeded to Dr. Mudd's . . . then to Cox's and there Booth was placed in the keeping of Jones. Great God, how my blood boils, and yet I admire the loyalty and fidelity of Jones. **His part was the*** *grandest of any that was played . . . nothing would tempt him! . . . Mrs. Surratt, Lewis Payne, George Atzerodt, and David Herold paid the penalty of the scaffold. Dr. Samuel Mudd, Michael O'Laughlin and Samuel Arnold were sentenced to imprisonment for life at the Dry Tortugas. Spangler got six years at the same place. Dr. Mudd was pardoned afterward [1869] and is now dead. John Surratt, who escaped to Italy, was brought back and tried . . . Corbett, who killed Booth, is in an insane asylum. And Jones is here to tell more than was ever known before!"* [69]

U. S. Post Office building at Bel Alton (Cox's Station) year 2000.

---

[68] The above excerpts are from an item in a c.1890 Washington, D. C. news journal titled *True to Wilkes Booth, A Man Who Would Not Betray Him . . . Face to Face With a Detective. The Man Who Ferried the Murderer Across the Potomac Meets Captain Williams, Who Led the Pursuit After the Memorable Ford's Theater Tragedy.* Name, date and page numbers for the clipping were left off the four-column magazine-type format item.

AFTER NOTE

The December 25, 1896 issue of the *Crescent* carried the obituary of Franklin A. Robey titled "The Last Survivor." The editor stated

> "... *the death of Mr. Franklin A. Robey removes the last of those who befriended John Wilkes Booth and aided him to cross the Potomac during his flight from Washington after the assassination of Lincoln. Dr. Samuel A. Mudd, Colonel Samuel Cox and Mr. Thomas A. Jones were intimately concerned in Booth's escape and have all been laid to rest. Mr. Robey was at that time overseer for Colonel Cox [at "Rich Hill"] and guided Booth and Herold from the latter's [Cox's] house to the place of concealment in the woods. During the days of hiding [April 16-21] the horses [of the fugitives] became restive and the fear that their neighing might bring about the discovery led to the decision to kill them. Mr. Robey knew of the location of certain quicksands and to these he and Herold rode the horses and shot them as they sank, bridled and saddled. A martingale ring was broken off in the passage through the undergrowth and this ring Mr. Robey had in his possession for a long time. Robey died at "Waverly" the 18th of December 1896 of typhoid fever at 65 years of age ... A successful farmer, highly respected and a good citizen, he was buried in the cemetery of the Methodist Church at Newtown in Charles County ... "*

Mood setting seldom is considered a legitimate technique of historiography. Yet, along with myth, legend and old wives' tales it may fit into a niche that helps history readers toward fuller understanding. Indeed, works of history cannot always be founded entirely on the bare bones of documentation. The poetic work included here was created by the late James Trent of the town of Seat Pleasant in southern Prince George's County, Maryland. For most of his life Mr. Trent carried on a love affair with southern Maryland ... its traditions, its land- and history-scapes and a wee bit of artistic fantasia that helped embellish the fascinating part of Maryland that adjoins the Nation's Capital between the Potomac and Patuxent Rivers. This "Lincoln 1865" work of Trent must be seen as the writer's understanding about 1864 conspiracy plans to kidnap the President and a return to Washington through Piscataway (about 15 miles north of Port Tobacco) after discussing the abduction with Charles County Confederacy sympathizers. Through most of the 19th century the two principal routes from Washington to southern Maryland were by way of Piscataway, or along a more easterly route through Surrattsville, T. B. and Brandywine. Booth's leg injury forced him to follow the easterly route from Navy Yard bridge because it would take him to the Dr. Mudd house more quickly.

Lincoln 1865.

Over these Charles County roads but sixty-five years before, a galloping servant had come to get a doctor for the dying Father of his Country, and Booth could see Dr. Brown's old house overlooking the Port Tobacco Creek, and as he approached Piscataway in the shades of evening he saw Mount Vernon across the wide water of the Potomac and fancied he could almost hear the steamers on the river toll their passing bells. Not the smallest idea entered his head that the man he meant to pursue would subdue the heart of the world by his love as Washington had done by his dignity.

As he passed out of the ruined town of Piscataway, whose old brick chimneys stood houseless like the widowers of many wives, he noted the long red brick Catholic Church with green shutters and yellow cupola, and under the wooden cross in its gable the words, "Come unto me all you that labor and are heavy laden and I will refresh you."

ADDENDUM A

GENEALOGICAL NOTES

Genealogists often are uncomfortable about tracing the lives of those having surnames as common as Jones. And early in the research for this work it was felt to be almost hopeless to travel through the past looking for the family of Thomas A. Jones. We were initially led to believe by records of the family left many years ago that the father of Thomas A. and a brother reached Maryland shortly after 1800 from a home in Wales. Federal Census findings (1800-1860) indicate this may be true. The Census reports did positively separate the T.A. family line from a much older Jones family here of considerable pedigree and wealth.

The key to identifying our Jones family lies in Federal Census figures for the Allen's Fresh District of Charles County, Maryland. During at least the first half of the 19th Century this district included many communities that lay some miles from each other: Cox's Station (Bel Alton), Allen's Fresh, Newburg, Faulkner, and Pope's Creek. All of these villages today lie along or near U.S. Highway 301, immediately north of the Potomac River bridge at Morgantown, which is directly across the river from Dahlgren in Virginia.

Decennial Census reports compared with Jones family items in 1844-1897 issues of the *Port Tobacco Times* reveal beyond a doubt that Thomas A. Jones was the oldest child of Elisha and Mary Jones who doubtless lived at "Rich Hill," the Cox family plantation that is today about four miles south of La Plata. The April 28, 1859 *Port Tobacco Times* announced the death of Elisha Jones at *"Rich Hill, his residence,"* April 16 at age 72. The 1840 Census for the Allen's Fresh District (2nd Election District of Charles County) shows nine people in the Elisha Jones family...five males and four females. Thomas A. would turn 20 in October, so he was shown as a male between 15 and 20. The 1850 Census was most revealing of all checked because both age and names are given. Now, Mrs. Elisha Jones is not included, and only children Ann Mary and Noble are listed. Thomas A. had married in 1845. The family now lived in the Hill Top District, about ten miles west of Rich Hill.

Establishing Ann Mary as a daughter of Elisha at the same time assured us that he was indeed the father of Thomas A. Census reports, Records Group 59 at National Archives, and a letter from the Little Sisters of the Poor sealed the previously conjectural relationship between Elisha and Thomas A. They were positively father and son.

A brother of Elisha said to have come here from Wales with Elisha very likely was the man whose death was reported in the February 26, 1851 *Port Tobacco Times*. He was Edward Jones and Elisha was executor of his estate . . . usually the responsibility of a respected relative or very close friend. And the *Times* announced the passing of another Edward Jones on July 5, 1855. Elisha served as Administrator in the settlement of the estate. This Jones, quite likely a nephew of Elisha, left considerable property . . . 317 acres along the eastern edge of the Mattawoman Swamp near Middletown. It appears that the two Edwards were father and son and therefore close to Thomas A. Jones.

The 1860 Census for Allen's Fresh shows the Thomas A. Jones family having eight children...a notable survival figure for those days. They were: Jane F., 13; Alice, 12; Richard, 8; Sarah E., 10; John James, 7; Edward C., 3; Henry C., 2; and Julia, 3 months. Mary Emma and Ernest E. were born later. This Census reported Thomas A. to be worth $8,000 in real estate and $7,000 in personal property . . . a very respectable total worth for any family at that time and place.

The following are being included primarily to allow now scattered descendants of Thomas A. and Jane Harbin Jones to find each other. However, contacts must be made through the author. Distance and time act to thin blood-tie relationships and some descendants of the T. A. Jones family are comfortable about their connections in a condition of anonymity for reasons known only to themselves.

Children of Thomas A. and Jane Harbin Jones . . . .

    Mary Catherine - 15 December 1845 - 28 January 1846

    Jane Frances   - 26 October 1846 - ?

    Alice Cecelia - 2 March 1848   - 26 December 1900

    Sarah Elizabeth- 1 October 1849 - ?

    Richard Thomas - 16 October 1851 - ?

    John James   - 16 July 1853   - 1908

    Ellen Rose   - 8 March 1855   - 12 October 1855

    Ann Carmilia   - 6 October 1856 - 10 November 1936

    Henry Clay   - 20 June 1858   - ?

    Mary Emma   - 30 April 1860   - 1938

    William Ernest - 26 December 1861 - 19 December 1942

Issue of Thomas A. Jones and second wife, Margaret Rountree...

    Edward Austin - 6 May 1880   - 25 February 1971
      born in Baltimore.

One son and a daughter of Edward A. Jones are living in northern Virginia, 1995: Roland and Helen Jones (Mrs. Paul) Horgan. Son Francis died in December 1995 and Wells many years earlier. He left a son Donald W. Jones, also in northern Virginia

My profound thanks go to these descendants of T. A. and Jane Harbin Jones for their cooperation and encouragement:

Miss Rita I. Cooksey (a granddaughter of John James)

Traci Penn (Mrs. Daryl E.) Rector (great granddaughter of Mary Emma) now living in Oregon

Raymond Anthony Jones (son of William Ernest) 1900-1992

Anna Jones Schellin (granddaughter of William E.) now living in Virginia. Mrs. Schellin's father was Raymond A. Jones.

Carolyn Jones (Mrs. Donald) Straeter (granddaughter of William E.) now living in California. Mrs. Straeter's father was Raymond A. Jones.

Helen Jones (Mrs. Paul) Horgan (daughter of Edward Austin, youngest child of Thomas A. Jones) now living in Virginia

Cora Irene Penn (Mrs. Robert) Quinn, (granddaughter of Mary Emma Jones 1860-1938). Mrs. Quinn knew two of Thomas and Jane Jones's children intimately during the first ten years of her life. The two Jones sisters were Ann Carmilia (1856-1936) and Mary Emma. Mrs. Quinn, her widowed mother, her grandmother Mary Emma and "Aunt Annie" lived near each other in southeast Washington in the 1930's. Sadly these two women imparted to Cora Irene no detailed family history. In fact, this branch of the Jones family, as with others contacted during research on this work, seems to have avoided discussions of their connections with Thomas A. Jones of Pope's Creek. There are some regrets about this now as the flight of time has slowly recast certain reflections of historic figures and events that touched the lives of the Thomas A. Jones family.

Any Jones family history owes credit to the late Mabel Angela Jones (daughter of William E. Jones) and sister of Raymond Anthony Jones. She acted as family historian in the 20th century. Mabel Angela was a member of the Sisters of Saint Joseph, All Saints Convent, in Baltimore, Maryland. She died in 1987, leaving family records and memorabilia to the two daughters of her brother.

[Author tape-recorded interviews with Miss Rita I. Cooksey, Mrs. Helen Jones Horgan and Mr. Raymond Anthony Jones. The originals of these interview tapes are in the archives of the Southern Maryland Studies and Research Center of the College of Southern Maryland, La Plata, Maryland 20646]

The Stuart Connection

The letter from the Little Sisters of the Poor about Ann Mary (Mrs. Hindle) Jones noted that her mother was Mary Stuart. Church records in Charles County often show nothing about early births, baptisms, marriages, deaths and burials. And many more people seem to have died intestate, leaving no record of kinship. In the case of a hinted-at Jones-Stuart family relationship the linkage was proven by a land record that recorded a property transfer executed March 28, 1842. The deed records the sale of land once possessed by Francis A. Stuart to Thomas A. Davis. All the survivors of Stuart are included since all were part of the agreement to sell the land...about eighteen acres, part of tracts called Part of Mayday and Part of Beech Neck. Elisha Jones and wife Mary (nee Stuart) were included among the names of ten family members, including the husbands of five married daughters, Mary among them. One son named Ignatius may have been a grandson of the Ignatius Stuart named in the Federal Census for 1790, which indicated he had a son under sixteen, perhaps Francis A., who had at least ten offspring in 1842.[69]

---

[69] Charles County, Maryland Land Records, Liber IB, Number 24, Folio 491-492.

ADDENDUM B

After 1844 lives of Thomas A. Jones family members were frequently reflected in pages of Charles County's one newspaper---the weekly *Port Tobacco Times and Charles County Advertiser* that was published in Port Tobacco.

JONES FAMILY NEWS ITEMS
(Printed in *Port Tobacco Times* 1845-1875)

Aug. 21, 1845, Elisha Jones delegate to county Whig convention from Coomes' District

May 13, 1847, Thomas A. Jones, constable, will auction three Negro women and children (no visible means of support). (Jones now a young man of 27.)

June 3, 1847, Thomas Jones, Whig from Allen's Fresh (doubtless Thomas A.)

June 24, 1847, Thomas A. Jones appointed constable in 5th district (included village of Allen's Fresh.)

June 22, 1848, Thomas A. Jones appointed by board of commissioners collector and constable for 2nd district (apparently left 5th district.)

May 29, 1849, Thomas A. Jones appointed collector for 2nd district

May 22, 1850, Thomas A. Jones delegate to Whig convention from Allen's Fresh district

July 17, 1850, Thomas A. Jones delegate from Allen's Fresh district to nominate four delegates to Whig Maryland State Constitutional Convention

Sep. 4, 1850, T.A.Jones, Collector, asks for all accounts to be settled

Feb. 12, 1851, Edward Jones died, 78 years, left widow and several children to mourn his death (possibly an uncle of T.A. Jones).

Feb. 26, 1851, Edward Jones, deceased, Orphans' Court. Elisha Jones, executor Edwin Jones, deceased, Orphans' Court. Robert S. Reeder, administrator

July 16, 1851, Thomas A. Jones candidate for constable from 2nd district

July 30, 1851, Thomas A. Jones attends Whig meeting at Allen's Fresh to choose delegates to convention at Port Tobacco

July 7, 1853, Thomas A. Jones attends Whig meeting and nominated as delegate to the Convention in Bladensburg (Jones now moving up in local Whig Party)

Oct. 27, 1853, Thomas A. Jones Whig candidate for justice of the peace in 2nd election district

Nov. 11, 1853, Thomas A. Jones elected justice of the peace

Nov. 30, 1854, Thomas A. Jones administrator for Bryan Robey, deceased, Orphans' Court

May 10, 1855, Thomas A. Jones serves on petit jury

July 5, 1855, Edward Jones, deceased, property to be sold, between Middletown and Mattawoman Swamp, about 10 miles from Port Tobacco, 317 acres. Elisha Jones, administrator

May 8, 1856, Thomas A. Jones serves on grand jury

June 11, 1857, Thomas A. Jones appointed one of twenty delegates to Democratic county convention in Port Tobacco

May 6, 1858, Thomas A. Jones serves on grand jury

June 24, 1858, Thomas A. Jones sells land part of his farm known as "Pope's Creek" 220 acres, on Potomac River nine miles from Port Tobacco, 2 ½ miles from Allen's Fresh. Steamboat landing adjoins land - two boats ply weekly to Baltimore and District

Mar. 3, 1859, Thomas A. Jones, administrator for estate of James Newman, deceased

April 28, 1859, Elisha Jones died at "Rich Hill," his residence, 16 April, 72 years old

Aug. 2, 1860, Lucinda M. Jones, deceased, estate probated, Elisha Jones, executor (Edward possibly an uncle of Thomas A.)

Dec. 20, 1860, T.A.Jones 4th Lt. in Allen's Fresh permanent organized rifle company

Jan. 3, 1861, Thomas A. Jones signs letter calling for a vote by the people of Charles County for or against a sovereign convention. 20 delegates from each election district would be chosen to meet at county convention and that body would elect three delegates to the state convention in Annapolis or Baltimore

Sept. 21, 1865, E.J.Jones goods sold at sheriff's sale to satisfy suits by John H. Cox, guardian Baalim Murdock; Alexander Penn, R.H.Mitchell and L.V.Oliver; James H. Grubb; William Reynolds, James R. McFarland and others (gives inventory)

May 4, 1868, Miss Ann M. Jones, dressmaker, in Port Tobacco, at Mrs. Stewart's, near the Court House (Ann Mary Jones now 30 years old)

Sep. 3, 1868, Thomas A. Jones election judge for 4th election district (Allen's Fresh)

Nov. 6, 1868, Thomas A. Jones insolvent estate (bankrupt but still active in public affairs)

Dec. 11, 1868, E.J.Jones (deceased) property on farm he occupied in Pomonkey Neck to be sold

Aug. 19, 1870, Thomas A. Jones judge of election

    Thomas A. Jones named in "Public Roads" list

Oct. 21, 1870, Thomas Jones name appears on corrected list of registered voters in 1st election district

    Thomas Jones name also appears on corrected list of registered voters in 2nd election district

July 21, 1871, Thomas A. Jones named in "Public Roads" list

June 7, 1872, Thomas A. Jones, Esq. constable in trial of a man accused of murdering a colored boy (homicide took place in Pope's Creek)

Nov. term, Thomas A. Jones served as grand juror

June 21, 1872, Thomas A. Jones, insolvent docket

Feb. term 1873, Thomas A. Jones was witness in state case

November Term 1873 Thomas Jones was witness in state case

Dec. 25, 1874, Thomas Jones name stricken off (removed) from voter registration list for second district

Oct. 1875 Thomas Jones name appears as a newly registered voter in second district

    Name of Thomas A. Jones disappears from local news items from 1876 until 1887, when he returned to the county to live near La Plata in the 1st election district.

[This roughly ten-year break in Charles County residency is a reflection of Jones living in Baltimore during his second experience with matrimony and patrimony]

May 5, 1893, *"Charles County at the World's Fair*

Mr. Thomas A. Jones has concluded to publish a book which will describe his connection with the escape of the assassin Booth to Virginia after he had shot President Lincoln. The book will contain a full account of the shooting and Booth's experience till he was shot in Virginia, and will also contain photographic views of places along the route, beginning at Ford's Theatre in Washington through Prince George's and this county to Pope's Creek. Mr. Jones will go to Chicago to sell the book at the World's Fair."

June 9, 1893, *"The enterprise that went out to Chicago to exhibit the Jones connection with the assassination of President Lincoln at the World's Fair, has, we learn, returned. It did not, we are informed, achieve the great financial success that its promoters had predicted. The exhibition, it is stated, failed to draw, and the book, which had been copyrighted, did not find ready sale."*

March 1, 1895, *"Justice of the peace, Thomas A. Jones is lying dangerously ill from a complication of diseases at his home at La Plata. Mr. Jones is seventy four years of age and his friends entertain but little hope of his recovery."*

March 22, 1895, *"Application was made by Col. L. A. Wilmer for letters of administration on the estate of Thomas A. Jones, deceased. Renunciation of John J. Jones filed."*

May 10, 1895, *"The Orphans' Court was in session on Tuesday with a full court. Letters of administration were granted to Col. L. A. Wilmer on the estate of the late Thomas A. Jones and warrant to appraise issued to John E. Mudd and W. H. Wenk."*

ADDENDUM C

OBITUARY

Thomas A. Jones

March 8, 1895 issue of

The Port Tobacco Times

"Death of Thomas A. Jones

Mr. Thomas A. Jones died at his home near LaPlata, on Saturday night last, aged seventy-four years. Mr. Jones had been in failing health for a long time, due to a general giving away of the vital powers, more than to any particular disease, and his death had been expected for several days. He was a native of this county and has resided here nearly all of his life. Mr. Jones was a man of strong character and was highly respected by everyone. In politics he was always a staunch Democrat and took an active interest in the success of his party. At one time he held a position on the police force of Baltimore city, and afterward an officer at the Maryland House of Correction. During the first Cleveland administration he was given a small place in the navy yard in Washington and held the position till the change of the administration. He then returned to the county and has for several years filled the position of Justice of the Peace in this district. His funeral took place on Tuesday, the interment being at Newport Catholic Church. The pall bearers were Messrs. J.H.Roberts, J.S.Turner, Col. L.A.Wilmer, R.F.Mattingly, J.E.Mudd and W.H.Wenk.

Mr. Jones is most widely known on account of the part he took in assisting Booth and Herold to escape to Virginia after the assassination of Mr. Lincoln. Mr. Jones was a foster brother of the late Col. Samuel Cox and when Booth applied to Col. Cox for assistance the latter, knowing well the kind of man Mr. Jones was, sent for him and put the fleeing criminal in his charge. Col. Cox and Mr. Jones were both suspected of knowing more about the escape of Booth from the county than they told, but so well did they keep their own counsel, and so well did Mr. Jones manage his part of the business, that though they were arrested and locked up in the old Capitol Prison in Washington, they were discharged for want of conclusive evidence. Mr. Jones first told his secret for publication in 1881, when it was published in full in *The Times*. In 1893 he published the whole narrative in book form and went to Chicago during the World's Fair to sell it but it found but little sale and he soon returned. Mr. Jones tells of the flight of Booth and Herold from Washington, their stop at the house of Dr. Samuel A. Mudd, *near Bryantown, on the morning after the assassination, and the setting of Booth's broken leg by Dr. Mudd, for which he was sent a prisoner to Dry Tortugas. He tells of Booth going to the house of Col. Cox, after leaving Dr. Mudd's, and how he [Jones] kept him hid in a pine thicket near where Bel Alton now is, and of the final escape to Virginia. During the time Booth was in hiding in the thicket, Mr. Jones visited this village daily to hear what the Federal troops were doing and on one occasion, while in the bar room of the Brawner Hotel, Captain Williams, a Federal officer, suspecting Mr. Jones knew more about Booth than he told, offered a reward of one hundred thousand dollars for information as to the whereabouts of the fugitive. Though Mr. Jones had lost nearly all of his property by the war, this opportunity to become rich did not for a moment influence him to betray the trust imposed in him, and he replied: 'That amount ought to get him if money will do it.'*

Mr. Jones was a strong sympathizer with the Confederacy, and in his book he gives an account of the part he took in assisting persons to get across the Potomac into Virginia and in forwarding the mails between the Confederate Government and the United States and Canada. He was appointed a member of the secret service of the Confederacy by Major William Norris, chief of the Confederate signal service, and the location of his farm at Pope's Creek, with his house on a high bluff, where he had a full view of the river for miles, both ways, was a great advantage to him in carrying on his secret work. A man named Ben Grimes, on the Virginia shore, co-operated with him in the dangerous undertaking, and they had a regular set of signals by which either could notify the other as to the condition of affairs and how to act. The river was patrolled by gun boats and a body of troops was stationed at Pope's Creek and another on Major Watson's place, less than three hundred yards from Mr. Jones's house. He tells how the signals were arranged and how boats crossed and recrossed the river, notwithstanding the vigilance of the Federal officers. He managed his part of the business with such discretion that he was never detected."

ADDENDUM D

The appointment of Rev. Lemuel Wilmer to Post Chaplain for the Post of Port Tobacco, Maryland
(signed by Edwin M. Stanton, Secretary of War)

WAR DEPARTMENT,
Washington, April 21, 1863.

Sir:

You are hereby informed that the President of the United States has appointed you a Post Chaplain, for the Post of Port Tobacco, Maryland in the service of the United States, to rank as such from the 21st day of April, one thousand eight hundred and sixty-three. ~~Should the Senate, at their next session, advise and consent thereto, you will be commissioned accordingly.~~

Immediately on receipt hereof, please to communicate to this Department, through the ADJUTANT GENERAL of the Army, your acceptance or non-acceptance; and, with your letter of acceptance, return the OATH herewith enclosed, properly filled up, SUBSCRIBED and ATTESTED, and report your AGE, BIRTHPLACE, and the STATE of which you were a permanent RESIDENT.

You will report for duty at the Post of Port Tobacco, Md. to the officer commanding there.

Edwin M. Stanton
Secretary of War.

Courtesy of Cynthia Q. Wilmer.
Original owned by the family of Ringgold Wilson Wilmer.

ADDENDUM E

Oath of Allegiance signed by Rev. Lemuel Wilmer

OATH OF ALLEGIANCE

I, _Rev. L. Wilmer_, Rector of Port Tobacco Parish, of _Charles_ County, Maryland, do solemnly swear that I will bear true faith, allegiance, and loyalty to the Government of the United States, that I will support and defend its constitution, laws, and supremacy against all enemies, whether domestic or foreign; any ordinance, resolution or law of any State Convention or Legislature to the contrary notwithstanding. Further, that I will not in any wise give aid or comfort to, or hold communication with any enemy of the Government, or any person who sustains or supports the so-called Confederate States; but will abstain from all business, dealing, or communication with such persons. And I do this freely, without any mental reservation or evasion whatsoever, with full purpose and resolution to observe the same; I also fully acknowledge the right of the Government to require this oath, the authority of the officer to administer it, and its binding force on me.

Subscribed and sworn to before me at Port Tobacco, Md. this 27th day of May, 1865.

(Signed) Rev. L. Wilmer

Lieut. and Provost Marshal

Courtesy of Cynthia Q. Wilmer.
Original owned by the family of Ringgold Wilson Wilmer.

ADDENDUM F

BIBLIOGRAPHY

Primary Sources

Charles County, Maryland, Bryantown/Mattawoman Church records, film #0013759, Hall of Records, State of Maryland, Annapolis, Maryland

Charles County, Maryland Land Records, Liber JST 1, Folio 354

Charles County, Maryland Land Records, Liber GWC, Folios 318, 319

Charles County, Maryland Land Records, Liber IB, Nr. 24, Folios 491-92

Charles County, Maryland Orphans' Court, 1895 October Term, signed by J. Benjamin Mattingly, Register of Wills

Charles County, Maryland Office of Register of Wills, *Inventories*, 1895

The Cincinnati Enquirer, Vol. XLI: No. 177: June 26, 1883, p 1, "GATH"

The Crescent, March 8, 1895, La Plata, Maryland.
    December 25, 1896, La Plata, Maryland.

CSA Maryland records, Maryland Historical Society, Baltimore, Maryland Thomas A. Jones Application for Membership in the Society of the Army and Navy of the Confederate States in the State of Maryland

George Alfred Townsend, 1883 diary entries, Maryland State Archives, Annapolis, Maryland

Index to City of Fort Smith, Arkansas death records, Book 2, pp.128 and 157

National Archives - II, College Park, Md., *General Records of the Department of State, Civil War Papers 1861-65*, "Records of Arrest for Disloyalty - 1861 and 1862." RG 59

The Planters' Advocate and Southern Maryland Advertiser, Upper Marlboro, MD, October 9, 1861

Port Tobacco Times and Charles County Advertiser,
    Vol. LIII, No. 37, February 12, 1897, Port Tobacco, Md.
    Vol. LIII, No. 49, May 7, 1897, Port Tobacco, Md.

Issues of February 26, 1851, April 18, 28, May 2, 1884, and March 8, 1895
Vol. XLIV, No. 15, Sept. 23, 1887

Provincial Residence, Little Sisters of the Poor, Baltimore, Maryland letter to authors, September 2, 1995

Taped interview with Mrs. Stephen Latchford in 1973 at her home in Mount Rainier, Maryland by John M. Wearmouth

Townsend, George Alfred telegram to T. A. Jones, undated

_____ letter to T. A. Jones, Sept. 18, 1883

_____ telegram to T. A. Jones, Oct. 21, 1883

_____ letter to T.A.Jones, Dec. 1, 1883

_____ letter to T.A.Jones, March 20, 1884

_____ letter to T.A.Jones, Apr. 8, 1884

_____ telegram to T.A.Jones, Apr. 12, 1884

True to Wilkes Booth, A Man Who Would Not Betray Him . . . Face to Face With a Detective. The Man Who Ferried the Murderer Across the Potomac Meets Captain Williams, Who Led the Pursuit After the Memorable Ford's Theater Tragedy. From a C.1890 Washington, D.C. publication. Name, date and page numbers for the clipping were left off the four-column magazine-type format item

U. S. Federal Government Decennial Censuses, Maryland, Charles County: Allen's Fresh and HillTop Districts, 1810, 1820, 1830, 1840, 1850 and 1860 (microfilm, La Plata Public Library, La Plata, Maryland, 1996)

War of the Rebellion Official Records, II-II, Washington D.C.: GPO, 1897.

Published Books

Baker, L. C. *History of the United States Secret Service*. Philadelphia: 1867.

Beymer, William Gilmore, *On Hazardous Service, Scouts and Spies of the North and South*, Harper & Brothers Publishers, New York, 1912.

The Equestrian Statue of Major General Joseph Hooker Erected and Dedicated by the Commonwealth of Massachusetts, Printed by Order of the Governor and Council, 1903, Wright & Potter Printing Company, Boston

Georgetown College Journal, Georgetown University, Washington, D. C.: April 1885 and February 1889 issues

Haynes, Martin A., *Second New Hampshire Vols.*, Manchester: 1865

_____ *A History of the Second Regiment, New Hampshire Volunteer Infantry in the War of the Rebellion*, Lakeport, N. H.: 1896

Hebert, Walter H., *Fighting Joe Hooker*, Chapter IV, "In Lower Maryland," Bobbs-Merrill: 1944

History of the Third Regiment, Excelsior Brigade: 72nd New York, York Volunteer Infantry, 1861-1865, Jamestown, N.Y.: 1902

Hutchinson, Gustavus B., *A Narrative of the Formation and Services of the Eleventh Massachusetts Volunteers from April 15, 1861 to July 14, 1865*, Boston: 1893

Jones, Thomas A., *J. Wilkes Booth*, Chicago, 1893

Kinchen, Oscar A., *Women Who Spied for the Blue and the Gray*, Dorrance & Company, Philadelphia: 1972

Leech, Margaret, *Reveille in Washington, 1860-1865*, Harper and Brothers, New York: 1941, pp 406-407

Oldroyd, Osborne, *Assassination of Abraham Lincoln*, Washington, D. C.: 1901

Pickerill, W. N., *Third Indiana Cavalry*, Indianapolis, 1906

Pogue, Robert E. T., *Old Maryland Landmarks*, Bushwood, Md.: 1972

Smoot, Richard Mitchell, *The Unwritten History of the Assassination of Abraham Lincoln*,
    Foreword by Orra Laville Stone, Clinton, Mass.: 1908

Taylor, Dr. Charles E., *The Signal and Secret Service of the Confederate States*, 1903

Townsend, George Alfred, *The Life, Crime, and Capture of John Wilkes Booth, and the Pursuit,
    ---------------------------Trial and Execution of his Accomplices*, Dick & Fitzgerald, Publishers,
    New York: 1865

---------------------------*Rustics in Rebellion, A Yankee Reporter on the Road to Richmond,
    1861-65*, University of North Carolina   Press, Chapel Hill: 1950

Wearmouth, John M., *Baltimore and Potomac Railroad - Pope's Creek Branch*, La Plata, Md.
    1986

_____*Charles County Helps Shape the Nation*, La Plata, Md.: 1986

Wearmouth, Roberta J., *Abstracts from the Port Tobacco Times and Charles County Advertiser*,
    Vol. II, 1855-1869, Heritage Books, Bowie, Md., 1991
    Vol. III, 1870-1875, Heritage Books, Bowie, Md., 1993
    Vol. IV, 1876-1884, Heritage Books, Bowie, Md., 1996

Periodicals

Century Magazine, April 1884

Blue & Gray Magazine, "The General's Tour - Booth's Escape Route: Lincoln's Assassin On the
    Run" by Michael W. Kauffman, Volume VII, Issue 5, June 1990

Surratt Courier, Vol. XII, No. 11, November 1987: Clinton, Md.

John Wilkes Booth

"HIS PART WAS THE GRANDEST OF ANY THAT WAS PLAYED."

Detective William Williams (of Thomas A. Jones)

This Thomas A. Jones work fills in many unknown aspects of the Booth-Herold escape account first exposed a century and a quarter ago. In late April 1865 journals coast to coast ran headlines about the assassins' flight following Lincoln's murder. And for decades following the United States press and authors end on end embellished and looked for new sensational angles to this story. And in the Washington, D.C. area a harbinger of spring each year is the hosted bus expedition along the Booth-Herold pathway through Southern Maryland. Now this Jones-Booth book gives as detailed an account as could be crafted from ten years of research of the Jones family, Charles County, Maryland, State and Federal Government primary and secondary resources. Southern Maryland pro-Confederacy leanings are an important part of the Jones contributions to the Southern cause. He never could have acted alone and successfully without much support from all levels of Southern Maryland society . . . as indicated in this fascinating tale. And "Tom Jones" was thought by Federal authorities in early 1862 to be an "extremely dangerous" Confederate agent appointed by Richmond to be the South's top secret service agent in Maryland.